Light Quanta: Einstein's Big Mistake

Ghanshyam Jadhav

Light Quanta: Einstein's Big Mistake

For permission requests, contact:
Ghanshyam Jadhav, Samarth Niwas, Omerga-413606, India
ghjadhav@rediffmail.com

Published by Ghanshyam Jadhav

Book Cover by Aditya Jadhav

Illustrations by Parameshwar Suryawanshi

First edition: 2023

ISBN 978-93-6039-195-9

Preface

It started in the year 1999. Everywhere there was curiosity about the arrival of the twenty-first century and at the same time there was a lot of fear of Y2K. At that time, I was joined Dr. Babasaheb Ambedkar Marathwada University, Aurangabad for Ph.D. work under the guidance of Dr. M. T. Teli for two years. Sir had given me one of his papers published in 1985 to study and now I had to work on it more. It consisted of the generalized Maxwell's equations to which an electric scalar field and a magnetic scalar field were added. The significance of those scalar fields was to be discovered. I began to study it carefully and realized that such scalar fields could only be created if the charge of the particle increased or decreased. The fields of non-conserving charged particles were drawn from it and taken to Sir for examination and I was relaxed. Because for all further work I would not face much difficulty. But sir didn't like the concept of non-conserving charge particle and so he rejected all my calculations. Actually, the question was about their scalar field and I couldn't do anything about it but it stopped my further work and I turned back to the study of electrodynamics. But now I got a new vision which was an attitude of curiosity and it came from finding out what those fields meant. At that time, along with superconductors, magnetic monopoles were also in discussion but they were nowhere to be found and could not be created in laboratories. So naturally I got curious about them. Actually, the first question that arose in my mind is that there is electric charge therefore there is an electric field, there is mass therefore there is a gravitational field, but there are no magnetic monopoles yet there is a magnetic field. Then I couldn't understand what exactly it was. A magnetic field exists due to the movement of electrons. The electron already has an electric field and due to its motion, there should be a change in the direction of the electric field, but it did not make sense to me that a new field was being created. There should an electric field but it seems that its some effect is not understood. Day and night I kept thinking that what is this magnetic field but I could not find any answer. I used to show colleagues from above that my Ph.D. work is going on but only the thoughts about magnetic field kept going on in my mind. My intellect was not letting me sit cheaply so I had a strong feeling that the magnetic field is not like what we think it is. There must be an electric field, but I don't know what its nature is. Many of my colleagues were working in ferrites and if I told them that magnetic field should not exist or no magnetic field, they would call me crazy, so I couldn't talk to them either. I wanted to try some experiments but I was not brave enough in front of others to tell them what I was doing. Then I had to wait till Sunday because not many people came to the department on that day and my guide also didn't come. Initially, I wanted to pass a current through a copper wire and check its field but at that time there was no battery so I used to remove my motorcycle battery and bring it to check the interaction between the wires. But all the signs seemed to be magnetic fields, so it was disappointing. A Cathode Ray Tube (CRT) was needed for further experiments but nothing could be done as other laboratories were closed. In the next two or three days, I used to collect other materials and keep them in my room and used to wait anxiously for the dawn of the next Sunday. Again, the following Sunday I would pass current through the wires and check the deflection with the CRT but there would be no indication of the electric field and I would get frustrated. I used to collect all the materials and put them back in the rack and go to the canteen and drink tea in frustration. This happened repeatedly. Well, I couldn't tell anyone about it, and the subject of magnetic fields never gone from my mind. One thought day and night, what would be the magnetic field. It seemed strongly that if I could just put a single electron in there and check its motion, I would get something, but that was impossible. CRT was the only option available and I checked the field of the coil using that, checked the field of the solenoid but all in vain. There was no indication of an electric field. When the direction of the CRT changed the direction of the force was changing and this was not an electric field sign. So, accepting that then there is only a magnetic field, I should have stopped but the mind was not ready to accept it. Sometimes I think that millions of people around the world have accepted the magnetic field. Maybe that's why they accept it as reality. But still my mind was not ready to accept it. Because of this, anxiety started to increase. Several weeks went by and gradually the thought of the magnetic field started to fade away from my head so that I became a little more stable. But still this question would pop up from time to time. And one evening suddenly I felt eureka-eureka and became so excited. I noticed that an electric field should be forming around the wire parallel to the wire and it should be applying an asymmetric electric force on the electron, so the electron is following a carved path. But there was no empirical evidence for this. That is, the static charge distribution could not produce a static asymmetric electric field, and the one produced by the steady current was assumed to be a magnetic field. Therefore, the matter was not going to be resolved. As such I realized that this could be used to solve the photoelectric effect but my question was the magnetic field. So, I did not give much importance to the photoelectric effect and that was my mistake. In fact, now the photoelectric effect has come to my rescue. That is, theorem stayed aside and corollary came to help. So, the conclusion was that the magnetic field is the effect of the asymmetric electric field, but as there was no empirical evidence, I ignored the matter and got out of it. But from time to time the matter would raise its head. Meanwhile, I was appointed as the Principal of Shri Chhatrapati Shivaji College, so naturally my administrative workload increased. But then in year 2020 the corona epidemic reared its head and the actual periods in college were stopped so there was a lot of time to think and the magnetic field reared its head again. But now my impatience was much less, so I could think calmly. At first, I thought that the photoelectric effect should be solved first and then the case of the magnetic field should be taken up.

A light wave is an electromagnetic wave and its electric field is asymmetric and I was realized that this is what causes the photoelectric effect. But when an asymmetric electric field applies a force on an electron, it does not make sense that the electron should follow a carved path. Because of the size of the electron, the intensity of the asymmetric electric field impinging on it was unknown. Then it seemed that the electron would follow the carved path only if the wave's electric field applied a force on the electron's electric field, but what is the actual evidence for this? In fact, we already have the empirical evidence, but we have ignored it. The force is mutual. That is, if the electric field of the wave applies a force on the charge of the electron, then the electric field of the electron must also apply a force on the charge accompanying the wave. But since there is no charge accompanying the wave, a force must be applied to the wave's electric field. That means the electric field of the electron must be applying a force on the electric field of the wave. This means that the wave's electric field must apply a force on the electron's electric field, i.e., the force must arise from field-field interaction. It's a wonder that so important thing has yet to be noticed. If this is the case, there is no point in blaming the classical mechanics. One thing came out of this is that as the asymmetry of the electric field increases, the kinetic energy of the photoelectron increases, and if the magnetic field is an effect of the asymmetric electric field, then if the frequency of the electromagnetic wave is increased, the magnetic force should be increased, which is outside the scope of conventional electrodynamics. Not only that, there should be a 90-degree phase difference between the electric field and the magnetic field in the wave, and both of these can be proven experimentally. This will answer what a magnetic field is and why magnetic monopoles are not found in the universe. This experiment will be an earthquake in science and will shake electrodynamics as well as quantum mechanics. Because this experiment will prove that light is a wave and has always been expressed as a wave, and Einstein's concept of 'light quanta' will slowly collapse. Now the next question arises that what is the length of a single light wave? For this, the first experiment that comes before the eye is Newton's Ring. In fact, the number of rings formed in this experiment should be equal to the number of crests and troughs in a single wave of monochromatic light used in it. As an alternative to this experiment, the air-wedge method can also be used in which bands of equal thickness are formed. I mean these experiments we have been using only to measure the wavelength or the thickness of the wire, but they are also telling how long a monochromatic wave can be but we never realized this. How many places in the world this experiment is done but it is a wonder why such important information has not come out yet. We do not know how we have become so blind. From this experiment it is observed that almost 600 to 700 bands are produced when sodium lamp is used. That is, due to the transaction of the electron in the sodium atom, which created this wave, the electron should vibrate at least 600 to 700 times at one place. So how can such an electron move in orbit? And if moving, how can it vibrate in one place at a time? Of course, it must not be moving in orbit. Then how can it be fixed without moving in the atoms? Suppose it is expressed as a wave in an atom according to quantum mechanics, how can one create a wave with about 600 to 700 crests and troughs? Again, we know that matter gets its magnetic properties from the spin motion of its unpaired electrons which is a physical motion and is only expected if the electron is a particle. If the electron in the atom is expressed as a wave, then there is no question of it having spin motion, so where does matter get its magnetic property? We don't know for sure whether electrons actually move around in an atom. We also don't know how the energy difference created by the transition of the electron turns into an electromagnetic wave. So why don't we put all this in front of the world and the new generation with an honest heart. Why are we hiding all this? Maybe the new generation will find the answer. How many fields can exist in the universe? Can there be any limits to it? No one talks about it. In fact, only one field can exist in the universe and if it is assumed to be an electric field then we have to find out what is the magnetic field and also what is the gravitational field. To be clear, it must be admitted that gravitational waves cannot exist. If the electric force arises from the field-field interaction, then light rays traveling close to an electron must be deflected. Also, the two light rays can interact and this all comes under classical mechanics. In fact, there cannot be separate mechanics such as classical or quantum. Again, we have to understand what is happening there and it is wrong to try to make new rules and form new equations for everyone. So, this is an attempt to put it all together. Some of these statements may be wrong or will become falls in future, but I'm sure there's a lot to take in. I am aware that I cannot express my thoughts in a coherent and logical manner. But I hope that the new generation will take right things from this and try to clear the confusions in the physics. These confusions have been going on for almost one and a half hundred years.

Many may wonder about the title of this book. But one thing to note is that in 1951, Einstein told his friend Basso that,

All these fifty years of conscious brooding have brought me no nearer to the answer to the question "What are light quanta?" Nowadays every Tom, Dick, and Harry thinks he knows it, but he is mistaken.

This phrase has been mentioned by many people in their lectures, research papers and books. I think they know the exact meaning. Einstein doubted that there would be nothing like the postulated light quanta. But people are cheering for light quanta now. He was probably worried that something else would turn up later,

showing that the light quanta were a trick used for the photoelectric effect. Because light is a wave and the energy in the wave cannot converge to a certain point, which he called light quanta. Secondly, he could not say what brings that energy together. What exactly that energy is, and why it depends on the frequency of the light wave, he could not tell. He was worried that it was just a trick used to solve the photoelectric effect. But people took the same thing to heart so much that a new mechanics emerged called 'Quantum mechanics' which is only leaps of imagination. Frankly speaking, it is difficult to get anything concrete out of it because its foundation is wrong. Another of Einstein's equations, which has received a lot of backlashes, is the mass-energy relation, $E = mc^2$, which is derived from the results of the Michelson-Morley experiment. In fact, the Michelson-Morley experiment was only intended to determine the true speed of light through the ether, a hypothetical medium for the transmission of light, as well as the true speed of the Earth. It made no observations about mass or energy. The conclusion of this experiment was that the relative speed of light could not be measured. But taking imaginary leaps from it, Einstein's train reached $E = mc^2$. This is called reaching heaven by a thread. Einstein's third case is the curvature of space. In fact, no one has yet understood what mass is, and no one has yet been able to arrive at what could be the root cause of the gravitational force. Gravitational waves have not been concretely discovered yet and will never be discovered. From the behavior of the gravitational force, it seems that a gravitational force should also exist between two parallel and closely spaced light rays. And people say that space is not filled with ether, it is a vacuum, so how can it be curved? The last 150 years have been nothing but confusion, and Einstein is also partly responsible for this confusion. While Einstein doubted the light quanta, he should have told the world that there was probably something else out there, look for it. Of course, we shouldn't have gone so far under his impression. But all this has happened.
Thank You.

Ghanshyam Jadhav
ghjadhav@rediffmail.com

"To my parents, teachers, friends and the universe, who gave me the gift of dreams and the ability to make them come true."

Contents

Chapter-01
Photoelectric Effect: The key point

1.1 History of Photoelectric effect

It was started in 1880 by German professor Heinrich Hertz. He was recently appointed as a professor at Karlsruhe University. Helmholtz, who was Hertz's doctoral supervisor there, expected Hertz to try for a prize held by the Berlin Academy of Sciences on problems in physics, but Hertz was not very interested but was impressed by Maxwell's proposals in electrodynamics. For this he proposed not as a challenge to Helmholtz but to Maxwell's proposition, can electromagnetic waves be transmitted in space? And can this be verified? And he started thinking about this and planned an experiment. He connected the antenna to a high voltage coil as shown in figure (1) and a series of sparks created high voltage in that coil so that high voltage pulses were generated in the antenna. At the same time, he made a loop of copper wire with two ends close together.

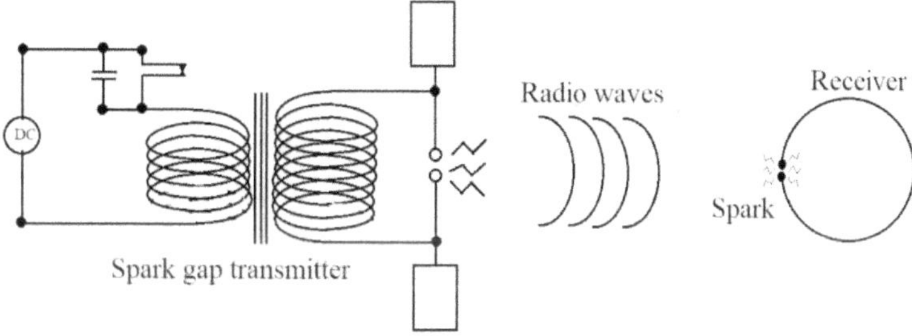

Fig. 1.1.1. Experiment used by Hertz to discover radio waves.

At the same time, he made a loop of copper wire with the two ends close together and placed the loop on the other side of the room. He found that when a series of sparks were produced in the first circuit, the copper loop also sparked. Sparks were being produced in the cooper loop though there was no direct connection between the first circuit and the loop. This means that a series of sparks from the first circuit was creating something and it was traveling through space to the loop placed on the other side of the room and creating these sparks, and they were, according to Maxwell, radio waves. Hence, at the age of 31, Hertz shot into the limelight overnight. But even then, he carried out the experiment very carefully in different ways and he found that the intensity of the spark was reduced when the shield was formed around the copper loop. Also, when a plain glass plate was placed in front of this copper loop, the intensity of the spark was reduced. He knew that ultraviolet light passes through a quartz plate, so when he placed the same plate in front of the copper loop, he found that the intensity of the sparks reversed, i.e., when the ultraviolet light was placed over the gap of the copper loop, the intensity of the sparks increased. This meant that the number of charged particles, we now call electrons, was increasing to cause a spark. That is, due to the ultraviolet light, electrons were released and this is the discovery of the photoelectric effect. Two physicists in particular drew attention to this, namely Sir JJ Thomson and Philipp Lenard. Further, Thomson discovered in 1899 that when ultraviolet light is thrown on a metal surface, the negative charges released are in the form of corpuscles, later called electrons, and are similar to cathode rays. Philip showed in 1902 that increasing the intensity of light increases the number of electrons emitted but does not change their energy. But as the frequency of light increases, the speed of the ejected electrons increases, but Philip could not explain why this happens, and the photoelectric effect remains a puzzle. Philip and some other physicists thought that the electrons are coming off the metal surface as soon as the light falling on the surface means, that is, the energy of the light does not change the energy of those electrons, but the electrons in those atoms already have a certain energy and the light is just triggering those electrons and those electrons are coming out with their original energy. But this explanation was not acceptable.

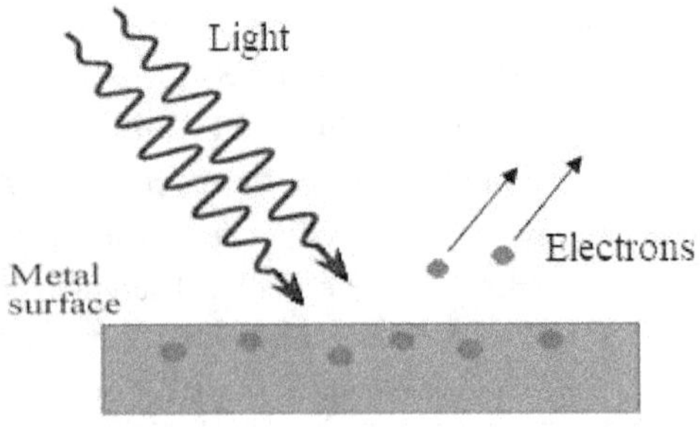

Figure 1.1.2. Photoelectric effect.

1.2 Difficulties faced by Classical Electrodynamics

Classical electrodynamics was fully developed in the 19th century and explained the mechanics of electricity and magnetism so completely and successfully that it became the subject of much discussion. Among these, the four differential equations proposed by Maxwell between 1855 and 1861, shown below, were very important.

$$\nabla \cdot \mathbf{E} = \frac{\rho}{\varepsilon_0}$$

$$\nabla \cdot \mathbf{B} = 0$$

$$\nabla \times \mathbf{E} = -\frac{\partial \mathbf{B}}{\partial t}$$

$$\nabla \times \mathbf{B} = \mu_0 \mathbf{j} + \frac{1}{c^2}\frac{\partial \mathbf{E}}{\partial t}$$

These equations are the essence of electricity and magnetism, i.e., electrodynamics. Any explanation regarding electricity and magnetism could be given using this equation, so this equation gained a very general importance. These equations show the direction and phase difference between the electric field and magnetic field in an electromagnetic wave propagating in space as follows.

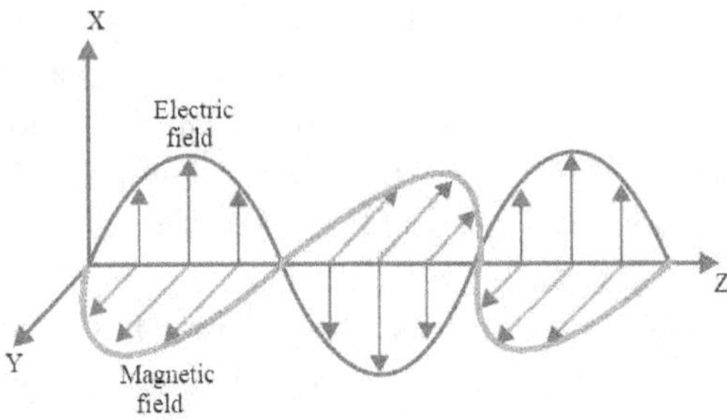

Figure 1.2.1. As the electromagnetic wave propagates along the Z-axis.

It is surprising that both the fields are perpendicular to each other and have zero phase difference, which means that both fields increase and decrease at the same time. However, the energy in it must be conserved. This means that when one field is increasing in energy, the other field must decrease and vice versa, and this must happen over and over again. But that doesn't seem to be happening. As this electromagnetic wave propagates through space with speed c, its energy also propagates with it. From this, in 1884 Poynting used Maxwell's equations to show how and in what direction the energy of an electromagnetic field can propagate through space. It is called a pointing vector, which is,

$$S = E \times H$$

This shows that the energy density of an electromagnetic wave propagating in space is,

$$u = \frac{1}{2}\left(\epsilon_0 E^2 + \frac{1}{\mu_0} B^2\right)$$

This shows that the energy in an electromagnetic wave depends on the electric field strength and the magnetic field strength in it. But light is also an electromagnetic wave so increasing the intensity of light increases the electric field strength and magnetic field strength in it. This means that increasing the intensity of light in the photoelectric effect should increase the kinetic energy of the electrons escaping from the metal but it was not happening and everyone was stuck here. This led to the conclusion that classical electrodynamics could not explain the photoelectric effect. Another important thing was that the effect was instantaneous. That is, when the light of a certain frequency falls on the metal surface, immediately the electrons will come out of the metal surface or will not come at all. If they didn't come out, no matter how long the light was on, they never came out. And this was surprising because in everyday life it was found that if energy is continuously given to a thing, it keeps collecting energy like if you start heating water it does not boil immediately but because of continuous heating it accumulates energy and as a result after some time the water starts boiling. Also, if the light is continuously thrown on the metal, the energy from that light should continue to accumulate in the metal, i.e., the electrons, and when additional energy is accumulated, the electrons should have come out of the metal, but it was not happening. On the contrary, the frequency of that light was increased and if it was equal to or higher than a certain frequency, the electrons were released immediately without any delay. So, Maxwell's equations or Poynting vectors i.e., electrodynamics could not explain anything. We are responsible for this because we have overlooked a few things, the first being that light has electric and magnetic fields and the electron is a charged particle, then ignoring how those fields can momentarily apply forces on the electron. And the second thing is to energize an electron in an electromagnetic wave without discussing what that process might be. Everyone stuck to the Poynting vector equation and concluded that classical electrodynamics could not explain photoelectric effect. The reason for this is the belief that Maxwell's four differential equations are the essence of electrodynamics and nothing outside of it. But if we do not know what the magnetic field is, such problems will arise. We will discuss this in detail later.

1.3 Einstein's explanation of photoelectric effect:

From the discovery of the photoelectric effect until 1905, when Einstein's 'light quanta' hypothesis was published, much work was done on it. All his properties were known and they were as follows.

1. In the photoelectric effect, the kinetic energy of the emitted electrons depends on the frequency of the light incident on the metal surface. The higher the frequency of the light, the higher the kinetic energy of the emitted electrons.

2. If the intensity of light increases, the number of emitted electrons increases but their kinetic energy does not change.

3. For every metal there is a certain frequency of light, if the frequency of incident light is less than that frequency then electrons are not emitted and if it is equal to or more then electrons are emitted. This frequency is called the threshold frequency of that metal.

4. The photoelectric effect is an instantaneous effect. When light shines on a metal surface, electrons are emitted immediately. It does not take any time.

All these properties could not be explained by classical electrodynamics. This story was read by young Albert Einstein. Until then there was another phenomenon that no one could explain blackbody radiation. Physicist Planck then proposed that atoms in a blackbody act as tiny oscillators and are coupled to each other. When they give energy to each other they give it in the form of a bundle, i.e., if an oscillator vibrates at frequency f it will give or take energy from the neighboring oscillator in the form of a bundle like zero or hf or $2\,hf$ or $3hf$ ...., where h is a constant, we now know as Planck's constant. That is, an oscillator in a blackbody could neither give nor receive energy in fractions of energy hf. Now there were bundles of energy formed and that is the quantum of energy and that is where the real quantum mechanics started. This proposition allowed Planck to explain blackbody radiation. Now one oscillator energizes another oscillator in the form of a bundle and it can be inferred that it must be electromagnetic radiation and light is also an electromagnetic radiation. So, in 1905, Einstein made a bold proposition that even though light is a wave, its energy is not dispersed but concentrated at certain points, called 'light quanta', with energy hf where h is again Planck's constant and f is the frequency of that light. Many now disagree whether Einstein's proposition depended on Planck's theory of blackbody radiation. A series of events at the time suggested that Einstein's proposition of light quanta was based on Planck's theory of blackbody radiation. But Einstein is nowhere to be found mentioning this. But there is no doubt that Einstein's proposition of light quanta was bold. But due to this proposition, the property of light as a wave has gone aside and the whole question has no meaning and at the same time there is no chance to ask any questions about the photoelectric effect. To be clear, Einstein's light quanta proposition was a compromise of the properties of light to explain the photoelectric effect. Since light is a wave and it is an electromagnetic wave, how can its energy converge to a certain point? And again, Einstein says that the energy of a light quanta is hf where f is the frequency of light, again light is a wave. In fact, no one at that time could explain photoelectric effect by assuming that light is an electromagnetic wave so there is no point in blaming Einstein. No one was accepting Einstein's explanation of light quanta but everyone was desperate. Einstein's explanation of the photoelectric effect was that energy in light is concentrated at certain points called light quanta and

In the photoelectric effect, one electron at a time absorbs the energy of one light quanta and part of it is spent for binding energy and the remaining part is converted into kinetic energy and the electron is released. Here it is assumed that since the electron is a point particle it can absorb the energy of a point particle, so light quanta, commonly known as photons since 1927, must also be a point particle. So, it may be assumed that an electron cannot absorb its full energy by interacting with a wave but can absorb its full energy by interacting with a photon. In fact, there are many questions hidden in Einstein's explanations that remain unanswered but have not been discussed much because light as an electromagnetic wave did not explain the photoelectric effect and there was no other alternative, so there was no point in questioning Einstein's explanation. Nevertheless, some questions raised by Einstein's explanation must be discussed. The first question is that even Einstein did not explain why the energy of the light quanta, we call photons, depends on the frequency, and no one has said anything since. Another thing is when Einstein says that electron is a point particle and light quanta is also a point particle so it can give energy to an electron. In fact, light is an electromagnetic wave and a single electron is responsible for producing it, and when it interacts with an electron in the photoelectric effect, there is no known verification of whether it interacts with the point of the electron or the field of the electron as illustrated in figure 1.3.1.

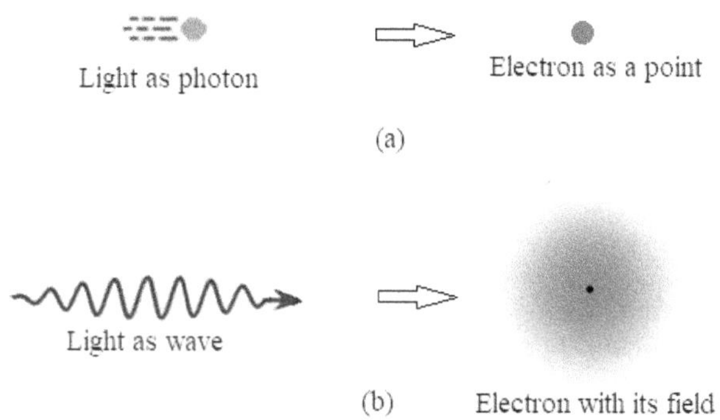

Figure 1.3.1. (a) Point photon interacting with a point electron, (b) Fields of electromagnetic wave (light) interacting with field of electron.

Basically, what is energy of light quanta? What is its physical form and how does an electron acquire that energy? Does he have some system or mechanism to capture that energy? And supposing that energy is taken, can it be a matter of what to do with it? Basically, the electron has no mechanism to absorb such energy and the energy has no physical form. It is a scientific idea of a man. Roughly we know two forms of energy namely kinetic energy and the other is potential energy. If an object is in motion, it remains in motion unless we stop it. So, we assume it has kinetic energy. Of course, there is no reason to say that there is any energy stored in that object. If an object is acted under a force, the tension created on the object due to that force is called potential energy. There is no such energy with that light quanta. If it has kinetic energy, that light quanta has no mass, and if it has potential energy, it has no force. Now if we consider an electron on that metal surface to move it from there another particle has to collide with it so that the kinetic energy of that particle gets to that electron and it gets out. Or the electron should be pulled out by applying an electric force, but no information is available about what is happening to the electron in the photoelectric effect. Basically, where our horses get stuck is on the idea that the energy of light is in its electric field and magnetic field. But we know how electric and magnetic fields apply forces on electrons, so why we didn't think about these forces is an unsolved puzzle. The important question here is whether the electric field and magnetic field of the wave can have energy or not? We know two forms of energy are kinetic energy and potential energy. It can be said that when the electric field in an electromagnetic wave is maximum, its tension or stress is maximum as shown in figure 1.3.2, then its potential energy must be maximum and kinetic energy must be zero.

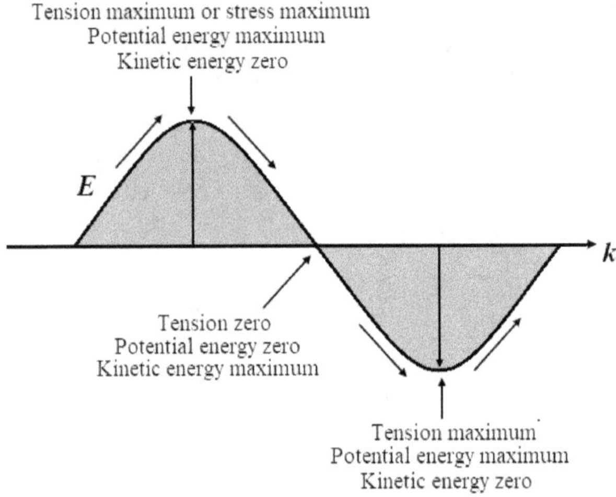

Figure 1.3.2. Potential energy and kinetic energy in electric field wave.

When the electric field tension decreases and its direction starts to change, the potential energy of the field decreases and the kinetic energy increases, and when the potential energy becomes absolutely zero, the kinetic energy should be maximum. Then again, the field tension will increase on the opposite side and when it becomes maximum, the kinetic energy of that field will be zero and the potential energy will be maximum. So, the question is what is a magnetic field? We will further discuss whether it could be the kinetic energy of the electric field. But the kinetic energy and potential energy of the field in the electromagnetic wave are not directly imparted to any charged particle but due to their potential difference a force is applied on the charged particle and thus it can be displaced. Therefore, the time line study that should have been done in accordance with how the electron should move with time after the electric field in the light interacted with the electron was not done, so the true explanation of the photoelectric effect could not come forward. And although late, Einstein's proposal of light quanta was accepted to explain the photoelectric effect, and he was awarded the Nobel Prize for it in 1921. But while awarding the Nobel Prize, the Royal Swedish Academy of Sciences got into a dilemma that the proposition of light quanta in the photoelectric effect was not accepted at all levels, and their Special Theory of Relativity was also not fully accepted, so finally the photoelectric effect was included in the awarding of the Nobel Prize. During that Nobel Prize speech, and again the following year, when Neill Bohr proposed the quantization theory of the hydrogen atom, he dismissed Einstein's light quantum concept and said that 'the hypothesis of light quantum is not able to throw light on nature of radiations'. Robert Milken of Chicago also tried hard between 1912 and 1915 to falsify Einstein's light quanta theory by empirical evidence but unfortunately, he did not succeed and instead of falsifying light quanta theory his efforts strengthened light quanta. In a research paper, published in 1916, he stated that Einstein's light quantum hypothesis is bold not too reckless of electromagnetic light corpuscles of energy hf. On the one hand he is saying that the well-established fact of interference is the universal truth that light is a wave, thus creating an interference pattern, but in the face of it, corpuscles of energy hf are being thrown into the light. First of all, this sentence is like surrender to the light quanta of Einstein. He was able to calculate Planck's proposed value of h in hf to within 0.5 percent from his experiments and was awarded the Nobel Prize in 1923 for his work and for measuring the charge on an electron. Although Einstein was awarded the Nobel Prize for the photoelectric effect in 1921, however, scientists were not ready to accept light quanta. During this period i.e., from 1920 to 1923 Compton was studying the behavior of X-rays. It was found to behave like a billiard ball when it hit an aluminum target. From that again the corpuscle nature of light was visible. That too was not acceptable to Neil Bohr. He designed his own experiment to disprove Compton, but he too discovered that X-rays behaved like billiard balls. So, most scientists had no choice but to accept Einstein's light quanta. Later, from 1926, the term photon for light quanta became common.

1.4. Consequences of the 'Light Quanta' Hypothesis

Light quanta were confirmed by Millikan's' experiment and Bohr's opposition was overturned. If light is a wave but manifests as a particle i.e., a photon, it is obvious that a particle, which is in motion, manifests as a wave. This was first proposed by de Broglie in 1924, saying that if a particle of matter, whether small or large, is in motion, it must have the properties of a wave. The wave accompanying that matter can be called a 'matter wave' and has a wavelength of h/p, where h is Planck's constant and p is the momentum of the particle. To be clear, this choice was a compromise because the energy of the photon is hf and the energy of the particle is mc^2, and how we can make both equal,

$$\text{photon energy} = \text{mass energy}$$

$$hf = mc^2 \quad \text{where c is the speed of ligh}$$

$$mc = hf/c$$

$$p = h/\lambda \quad \text{where p is the momentum of photon and}$$
$$\lambda \text{ is the wavelength}$$

$$\text{or}$$

$$\lambda = h/p$$

By collecting up such things, the equation for matter wavelength proposed by de Broglie, which was for photos, is now believed to be applicable to matter particles. Thus, the wave properties were applied to the matter and a new path was created for further travel. The next question arises as to whether a particle in motion can be expressed as a wave, so without assuming a continuous and long matter wave, as shown in figure 1.4.1, a wave packet formed by the interference of multiple waves is assumed.

Figure 1.4.1. de Broglie wave packets.

Because assuming a long chain of matter wave, one could not tell exactly where the particle is. Of course, depending on whether the wave packet is short or long, we started talking about the possibility of finding the particle exactly.

Another thing is that the matter wave is scalar wave or vector wave. When a cannonball is moving at high speed its wave can be expressed as shown in figure 1.4.2. But de Broglie does not give any explanation as to whether this wave is a scalar wave or a vector wave. Light is a vector wave but it is expressed as a particle according to Einstein, so de Broglie's matter wave can be a scalar and there is no explanation.

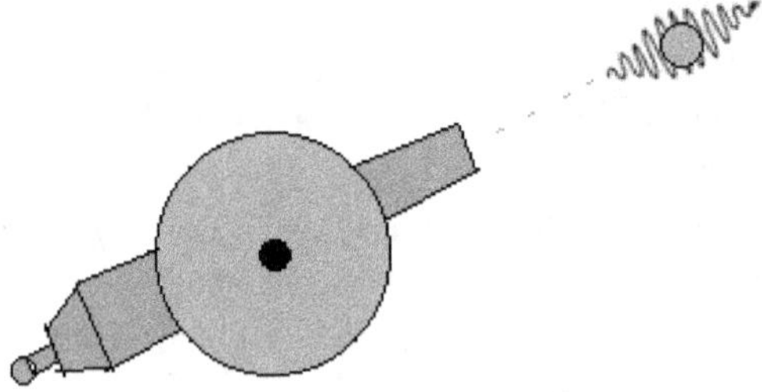

Figure 1.4.2. De Broglie wave packet associated with a cannonball.

It is understood that in the case of a short-wave packet, the position of the particle can be determined accurately and in the case of a long wave packet, the momentum of the particle can be determined accurately. It means that if one thing is compromised, one has to compromise with the possible thing arising from it. From this Heisenberg formulated the Uncertainty Principle, which states that an attempt to accurately measure a particle's position will increase the error in attempting to measure its momentum at that time, and vice versa. From this the Uncertainty Principal was also proposed in energy and time and some other. This confirmed that nothing physical could be evaluated strictly, and all these results were the matter wave proposed by de Broglie which started from Einstein's light quantum theory. Further, Schrödinger proposed the equation of motion by assuming the wave nature of the particle instead of assuming its corpuscular nature, the equation is called Schrödinger equation and from that equation the behavior of the particle was predicted, from which the energy of the particle was seen to be quantized. The so-called approach to matter opened up numerous research opportunities for physicists. So, no one ever questioned whether we needed to look back to Einstein's light quanta, which started it all. Rather, all post-1905 theory rests on Einstein's light quanta hypothesis. So, its contributors have no reason to doubt Einstein's hypothesis of light quanta. Conversely, if this assumption is proven wrong, the 125-year-old superstructure could collapse. If we look closely at what exactly this quantum mechanics gave us, we cannot say anything concrete. Only uncertainty is given. The approach to this subject has become more complex and complicated. The simple fact is that we don't even know exactly how the light waves that we perceive in the universe are formed. That is, we understand that after an electron in an atom jumps from higher energy level E_1 to lower energy level E_2, the energy difference E_1-E_2 is emitted in the form of an electromagnetic wave whose frequency is $f = (E_1$-$E_2)/h$. It is not known how that energy difference is converted into an electromagnetic wave. If we say that the jumped electron vibrates with frequency f in that atom. But this is not possible because if the Rutherford-Bohr atomic model is assumed, electrons are moving in orbits, and if the cloud model of the atom is assumed, the electrons are expressed as waves. There is no possibility of electron vibration in both, so we still do not know how an electromagnetic wave of frequency f should be generated. It is not known whether electrons move reliably in orbits in the atoms that make up our bodies. In the last 125 years, we have not known the material things surrounding many such beings and we have been unforgivably ignoring them. The root cause of all this is the photoelectric effect, and solving it requires removing the mantle of quantum mechanics that has evolved over the last 125 years over our brains. So, we have to go back to the horizon of 1905 and settle this question.

1.5 Explanation of photoelectric effect in terms of light as an electromagnetic wave

To think of explaining photoelectricity by assuming that light is an electromagnetic wave is beyond imagination at the present time, and even if one succeeds in explaining it, no one is likely to pay attention to it. But no matter how we explain it and how much compromise we make, it does not make any difference to what happens, and that thing happens according to its principle and there is no change in it with time. So, no matter how much time passes and if there is a mistake made in understanding something, sometimes the truth comes out. The same thing happened with Einstein's light quanta hypothesis. Because this hypothesis was not accepted for many years. This meant that there must be something wrong, but it was not clear. Now, even though everyone is jumping on it, if one's intellect does not agree with it, he should try to trace and solve it with the necessary logic. Then we are going to try to track step by step what is happening at that exact place in the photoelectric effect. As shown in Figure 1.5.1, when a light beam of a certain frequency is incident on a metal surface, electrons are ejected, which is called the photoelectric effect. The condition is that the frequency of the light should be equal to or higher than the threshold frequency of the metal. The emitted electrons are called photoelectrons. We know that a light wave is an electromagnetic wave. This means that when it hits a metal surface, it will apply a force on the electrons there. To understand this, a light wave of wavelength l which is an electromagnetic wave is incident on a metal surface as shown in figure 1.5.2, and applying a force on the electrons there. Initially we will consider only the electric field in the electromagnetic wave. Since the electron has a negative charge, the electric field in the wave will apply a force on the electron in the opposite direction to the field whose equation is $F = -eE$, where F is the force on the electron, e, charge on the electron and E, the electric field at the position of the electron as shown in figure 1.5.1. Here a question arises that the electron is a point particle and how much electric field of the wave has to link to it so that the force results. Another thing is that the electric field of the wave around the electron is not symmetric as shown in the figure, so will it affect the force? and will the electron go straight in the direction of the force or in a curved direction? This must be understood. We know that the force is mutual, that is, if the electric field of the wave applies a force on the charge of the electron, then the electric field of the electron must apply a force on the charge with the wave electric field. But the wave's electric field is accompanied by no charge that must be responsible for creating that electric field. But mutual force must exist. That means the electric field of the electron must be applying a force on the electric field of the wave. If this is true, it means that the wave's electric field will certainly be applying a force on the electron's field rather than applying a force on the electron's charge. So, our belief that an electric field applies a force on a charge is wrong. In fact, even the concept of charge becomes meaningless. What matters is the strength of its own electric field around that particle. Now it is not known what caused him to acquire this field. So, the concept of electric field which was assumed to understand the force between two charged particles is not imaginary but real. And the electric force between them is not an action at a distance but a force is created between them as their fields interact with each other as shown in figure 1.5.2. This requires a generalization of the electric force equation $F = qE$ through field-field interaction to form a new equation that can be expanded in the form of ($F = E_1 * E_2$), and this thing has remained unchanged. A detailed discussion of when two fundamental charged particles can attract and when they repel is given in the chapter Fundamental Particles, Fields and Waves. At present we know enough that electric force arises from field-field interactions. However, how this generalization is necessary and true can be analyzed step by step as follows.

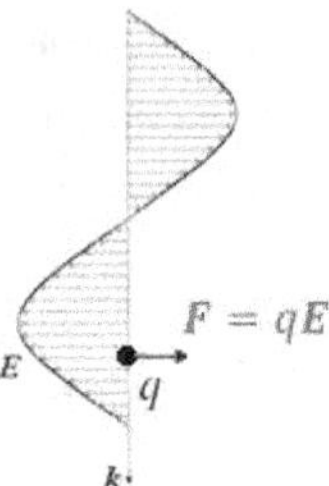

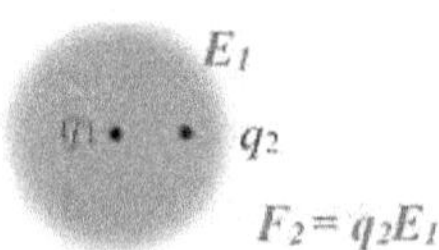

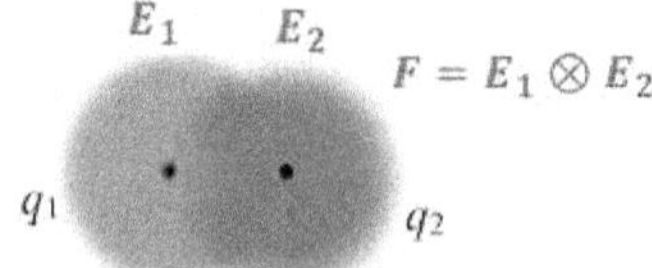

(a) Force through Charge-Charge Interaction

(c) Force on the chrge q due to electric field E of EM wave. Force is mutual. Electric field of charge q will also apply force on the electric field E.

(b) Fource through Field-Charge Interaction

(d) Force through Field-Field Interaction

Figure 1.5.2. Generalization of electric force.

Case (1): In fig. 1.5.2.(a), the force between the two charges is expressed in terms of their charges. It may be considered as force in terms of charge-charge interaction.

Case (2): In fig 1.5.2.(b), charge q_1 produces electric field E_1 around itself. If another charge q2 is brought into the field then the field applies force on the charge. It may be considered as force in terms of field-charge interaction. The force equation is given by $F_2 = q_2E_1$, where E_1 is the electric field produced by charge q_1 at position of charge q_2. Force is mutual. Therefore, charge q_2 also applies force on charge q_1 in terms of field-charge interaction by $F_1 = q_1E_2$, where E_2 is the electric field produced by the charge q_2 at position of charge q_1.

Case (3): In fig. 1.5.2.(c), an electric field wave in an electromagnetic wave is propagating with speed of light and interacts with a charge q at an instant. This field also applies force on the charge in terms of field-charge interaction. If E is the electric field of the electric wave at position of the charge q at any instant then the force equation is $F = qE$. But the force is mutual. Hence the charge q should also apply force on the electric field of the wave. As there is no charge associated with electric field of the wave, the electric field of the charge q must apply force on the electric field of the wave which may be considered as force through field-field interactions. Obviously, as the subjected charge q is always associated with its own electric field, therefore, the electric field of the applied wave must apply force on the field of the charge resulting into the force through field-field interactions. Thus, finally one can conclude that the electric force exists not in terms of charge-charge interaction or field-charge interaction but it exists in terms of field-field interactions.

Case (4): In fig. 1.5.2.(d), if two charges are brought close to each other, then force between them, either attractive or repulsive, is existed in terms of their field-field interactions. It means the electric field of first charge exerts force on the electric field of the second charge and vice versa. Thus, a new electric force

10

equation is to be developed which could explain the force in terms of the field-field interaction. At present we consider existence of electric force in terms of the field-field interactions for further discussion.

A light wave of amplitude l incident on a metal surface is shown in Figure 1.5.3 showing the wave of the electric field in the x-z plane. Metals have free electrons but they cannot come easily out of the surface. As shown in Figure 1.5.4 (a), when the electric field of the light interacts with a point A to the metal surface, the value of the field is zero, so no force will be applied on the electron at that point, so no effect will be observed. As the light wave travels, when the electric field between sections A and B interacts with an electron at that point on the metal surface, that electron will be pushed in the opposite direction of the electric field. At this time, the electric field of the light will interact with the electric field of the electron, and since the effective electric field of both is relatively the same size, the electric field of the light may interact with only one electron at a time. Of course, as shown in Figure 1.5.4 (b), here the electric field of the light is asymmetric around the electron, so it will apply an asymmetric electric force on the electron. To understand that effect, we divide the electron's electric field into two hemispheres, with electric field A in the lower hemisphere and electric field B in the upper hemisphere. Suppose the electric field of the light exerts a force F_A on A and a force F_B on B, surely F_A will be smaller than F_B. Both the forces are going to be in the same direction and opposite to the electric field, but since the Fa force is greater than the Fa force, the electron will be pushed in a curve instead of in a straight line, i.e. inside the metal. Therefore, when the electric field between the sections of light A and B interacts with the metal surface, the electrons will not be able to escape from the metal surface, so there will be no photoelectric effect.

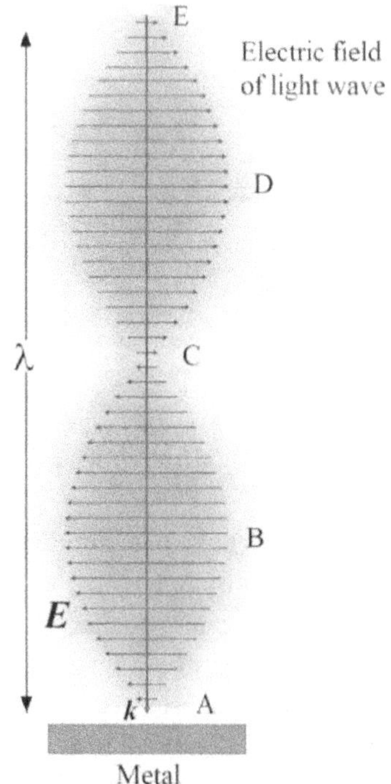

Figure 1.5.3 Electric field in a light wave of wavelength l incident on a metal surface.

11

As shown in figure 1.5.4 (c), when the electric field of the light at point B interacts with the electron on the metal surface, the electric field of the light around the electron will be symmetric, so both the forces F_A and F_B will be equal, so the electron will be pushed in a straight line i.e., parallel to the surface. This means that the photoelectric effect will not be seen here either.

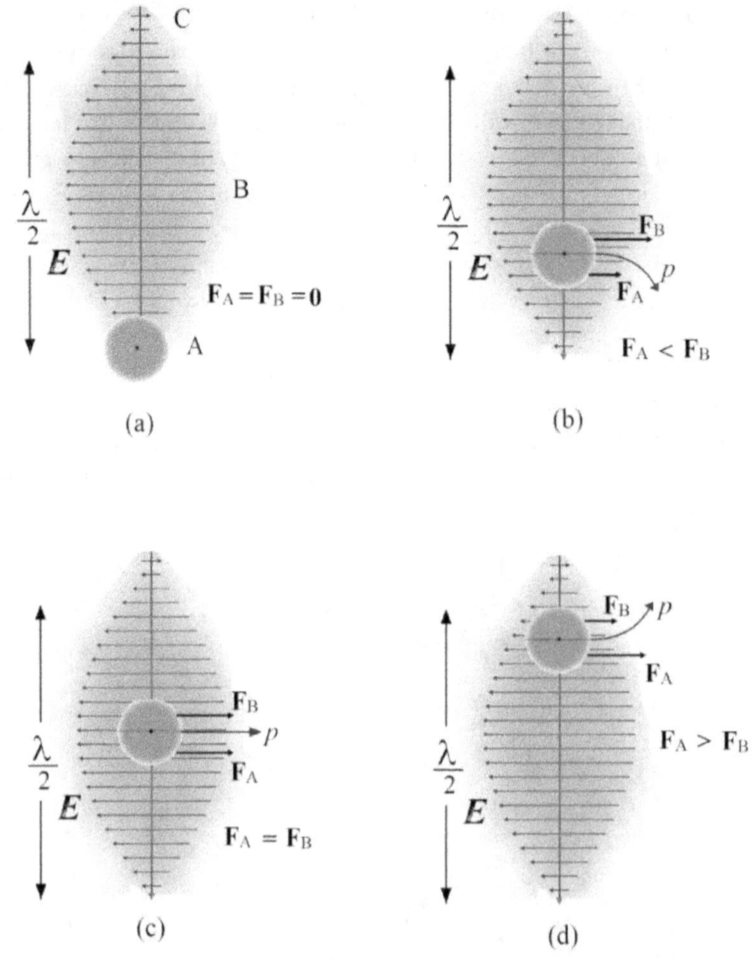

Figure 1.5.4. Electron interacts with electric fields in different sections of light wave.

As shown in Fig. 1.5.4 (d), when the electric field between the B and C sections of the light is linked to the electrons on the metal surface, the force F_A will obviously be greater than the force F_B. So instead of pushing the electron in a straight line, it will be pushed in a curved line and because of the strong force above the electron's lower field, it will be pulled out of the metal surface. This means that photoelectric effect can be seen here. How high it will be pulled will depend on the difference between these two forces. Since the electron moves in a curved line, we can assume two effective forces on the electron as shown in Figure 1.5.5, a force **F** which is the effective electric force in the direction opposite to the electric field and the other apparent force Fa pulling the electron up and which is created by the difference between FA and FB. That is,

$$\mathbf{F_a} \; \alpha \; |\mathbf{F_A} - \mathbf{F_B}|$$

If the electron does not move, Fa will be zero. This means that here Fa will also depend on the velocity of the electron. That is, the higher the speed of the electron, the greater the Fa. Another thing to mention is that this apparent force Fa must be perpendicular to the velocity of the electron and in the plane of the electric field.

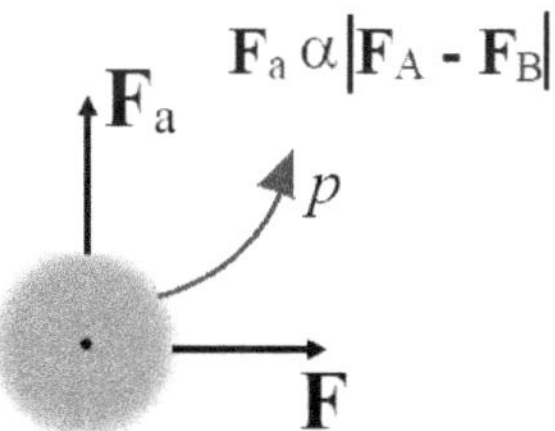

p is the path followed by
the electron.

Figure 1.5.5. When an electron is under the influence of two forces, the effective electric force F and the apparent force $\mathbf{F_a}$.

The total work done to pull the electron out of the metal surface in this process will depend on the effective electric force F that accelerates the electron i.e., the electric field of the wave and how much that apparent force Fa pulls the electron will depend on how much the difference between F_A and F_B is, or how asymmetric the electric field is, or how short the wavelength is. That is, the work done by the electric field between sections B and C of the light must be proportional to 4E0/l to pull the electron out of the metal surface. That is,

$$\text{Work done } \alpha \; \frac{4E_0}{\lambda}$$

$$\text{Work done } \alpha \; v E_0$$

where v is the frequency of the light wave.

$$\text{Work done} = k v E_0$$

where k is the proportionality constant.

Some of this work will be used to free the electron from the surface and the rest will be converted into kinetic energy of the electron. That is,

$$k v E_0 = W_0 + \frac{1}{2} m v^2 \tag{1}$$

This clearly shows that the kinetic energy of the ejected electron depends on the frequency n of the light and also on amplitude E_0 i.e., the intensity of the light wave. But empirical evidence suggests that kinetic energy does not depend on light intensity. That is, it shows that the electric field of a single light wave is interacting with the electric field of one electron at a time and its amplitude E_0 is constant. Because that light wave is created from single transition electron and E_0 is the maximum value that corresponds to the maximum kinetic energy of the electron. The equation for the photoelectric effect given by Einstein using the light quanta hypothesis is,

$$h\nu = W_0 + \frac{1}{2} mv^2$$

(2)

This means that the value of Planck's constant, h, must be related to the effective electric field of the electron, on which the electric field strength or amplitude E_0 of the light wave it produces will depend.

1.6 Experimental evidence for the photoelectric effect

In fact, the photoelectric effect is the empirical evidence for equation (1.5.1) but since it has been used by Einstein's light quanta hypothesis it remains to be seen whether any other evidence can be found. Equation (1) actually includes E_0, but since the photoelectric effect occurs for frequencies in the light wave frequency range or higher, we cannot change E_0. Therefore, its effect on the kinetic energy of the electron in the photoelectric effect is not observed. Another thing we should note is that we have not yet considered the effect of the magnetic field in the light wave and for this we need to understand what the apparent force F_a is. This force is generated in an asymmetric electric field whose curl is non-zero. Again, it depends on the velocity of the charged particle and it changes the direction of velocity of the particle. This indicates that it must be acting as a magnetic force. That is, a magnetic field can be an effect of an asymmetric electric field, so the magnetic field must not have an independent existence. There are no magnetic monopoles in the universe, and a single field must exist in the universe, and if it is an electric field, then the magnetic field and the gravitational field must not exist independently. We will discuss this in detail in the chapter 'Ether, Particles, Fields and Waves'. If the effect of asymmetric electric field is the magnetic field, it should be demonstrable. But unfortunately, a static asymmetric electric field cannot be created due to static electric charge distribution and whatever static asymmetric electric field is created, we already perceive it as magnetic field which is created due to uniform motion of electric charge i.e., due to steady current. But in a light wave the apparent force F_a is frequency dependent and if it acts as a magnetic force then increasing the frequency of any electromagnetic wave must increase the applied magnetic force. An experiment can be designed to test this.

For experimental verification, an electromagnetic wave produced by an antenna, without directly using light waves, can be used and a cathode ray tube (CRT) can be used to measure the strength of its electric and magnetic fields. Suppose an electromagnetic wave of a certain frequency is being generated by a vertical antenna as shown in figure 1.6.1 which will propagate around it at the speed of light i.e., c. We will measure the strength of the wave as shown in the figure propagating along the z-axis. For that, two CRTs have to be used, one to measure the electric field strength and the other to measure the magnetic field strength. To measure the strength of the electric field at a certain distance from the antenna, a CRT let's say (CRTx) has to be placed parallel to the X-axis as shown in the figure. The electron beam from this antenna will oscillate up and down with the applied force in the electric field in the wave creating a vertical line on the screen of the CRTx. The greater the strength of the electric field, the longer the line will be. By varying the alternating current supplied to the antenna, the strength of the electric field can be varied and the effect will be reflected in the line length produced by the CRTx on the screen, i.e., on the deflection of the electron beam. For that,

14

the dimensions of the CRT can be used to determine the frequency of the electromagnetic wave to be generated. Suppose the electron beam length in CRT is x and the final accelerating anode voltage is V, then the kinetic energy of the electron is

$$\frac{1}{2} mv^2 = eV$$

From this the velocity v of the electron can be calculated.

If T is the time taken by an electron to reach the screen after leaving the electron gun, it can be calculated as,

$$T = \frac{X}{V}$$

The time period, T', of the electromagnetic wave produced by the antenna should be at least twenty times T, ie, T' = 20T. So, the maximum frequency of that wave should be F = 1/T'. This means that electromagnetic waves of maximum frequency F can be used for that CRT. Investigation is to be done by varying frequency of the electromagnetic wave; one should start with a lower frequency. Since the deflection of CRTx depends on the strength of the electric field and not on the frequency of the wave, even if the frequency is changed, the deflection of the CRTx should not change. So, this means that when the frequency of the electromagnetic wave produced by the antenna is changed, its electric field strength remains constant which is necessary for the verification of our experiment.

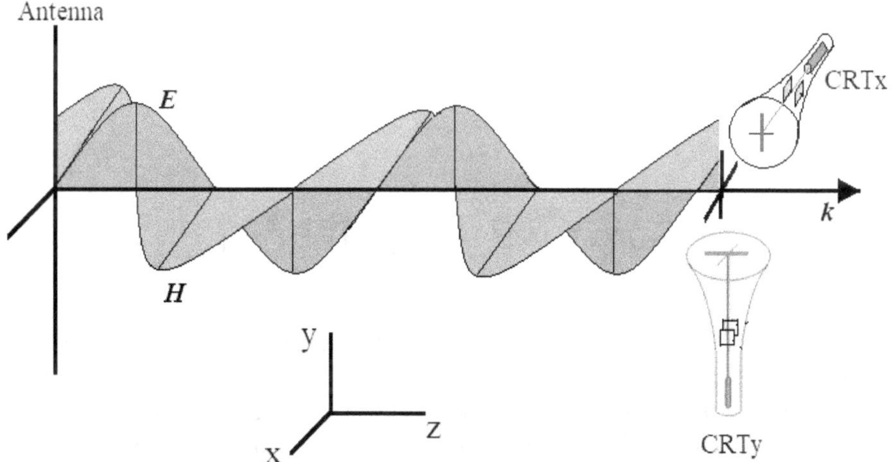

Figure 1.6.1. An experimental design for measuring the strength of electric and magnetic feeds in an electromagnetic wave produced by an antenna.

Now to check the strength of the magnetic field in that electromagnetic wave, another CRT, called CRTy, has to be placed parallel to the Y-axis i.e., perpendicular to the magnetic field. According to conventional electrodynamics, the magnetic field in the wave has no effect on the electron beam in CRTx. But the electron beam in the CRTy will continue to be deflected to the right and left and will create a horizontal line across the screen whose length will depend on the strength of the magnetic field in the wave. Specially, there should be no change in the deflection of the CRTy when the frequency is changed by keeping the strength of the magnetic field constant. Because according to prevailing electrodynamics

$$\mathbf{F} = q(\mathbf{v} \times \mathbf{B})$$

is the magnetic force equation. In this, B, the magnetic induction, is the maximum value of the magnetic field in the wave that is proportional to the amplitude of the wave and is constant and v is the velocity of the electrons in the electron beam. Hence the deflection of its corresponding beam will be maximum and remain the same forever. But if the magnetic field is the effect of an asymmetric electric field, the apparent force Fa will act as a magnetic force which will increase as the asymmetry of the electric field increases i.e., as the frequency of the wave increases. That is, by keeping the strength of the electromagnetic wave constant, if the frequency is increased, the deflection of the electron beam in the CRTy should increase, which can be understood as follows.

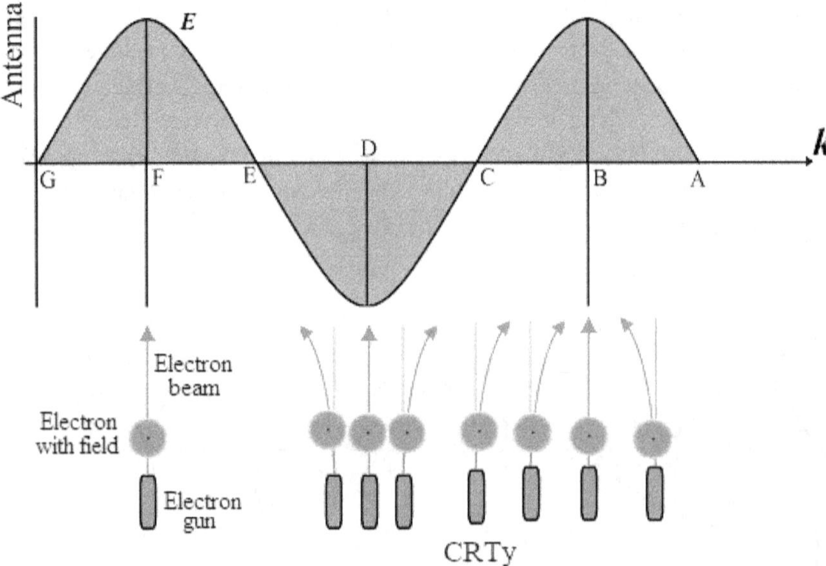

Figure 1.6.2. Deflection of electron beam of CRT at different sections in electromagnetic wave.
Figure 1.6.2 shows the CRTy in its different sections assuming only the electric field at a particular time in that electromagnetic wave. In sections A and B the wave's electric field is positioned to accelerate the electrons in the CRT's electron beam. The force it applies arises from field-field interactions in which the electric field of the wave applies a force on the electric field of the electron and the electric field of the electron applies a force on the electric field of the wave. The wave's electric field is increasing as it moves from point A to point B, meaning it is asymmetric around the electron. To understand this effect, the electron's electric field can be divided into two hemispheres, A on the right and B on the left. The wave's electric field is weak in hemisphere A and strong in hemisphere B, so the force F_B it exerts on hemisphere B will be stronger than the force F_A exerted by the wave's electric field on hemisphere A. These forces oppose the electron's velocity so the electron will decelerate, i.e., slow down. As a result, hemisphere B will slow down more than hemisphere A so the electron will be pushed towards point B of the wave, i.e., towards where the electric field is strong. When the CRTy is at point B of the wave, there will be equal force on both hemispheres of the electron, so both will decelerate by the same amount, so the electron will move in a straight line, i.e.,

towards point B. When the CRTy is between section C and section B of the wave, the electron hemisphere A has a stronger force than hemisphere B, so the electron is deflected back to point B. When the CRTy is at the C point of the wave, the electron beam will decelerate the electric field linked to the electron hemisphere A and at the same time the electric field of the wave linked to the hemisphere B will accelerate so that the electron will return to the B point. When the CRTy is in sections C and D of the wave, the wave's electric field will cause the electrons to accelerate, i.e., increase in speed. The electric field of the wave above the hemisphere B is stronger than the electron's hemisphere A, so the velocity of the hemisphere A will increase faster than that of the hemisphere B, thus the electron will be pushed towards the C point i.e., towards the weaker field. Thus, the deflections of the remaining electron beam in Figure 1.6.2 can be explained. Now this wave is moving forward with time at a speed c so the CRTy will also sequentially link the electric field in all sections of that wave and deflect the electron beam according to the electric field in that section as shown in the figure. As a result, a horizontal line formed by the electron beam will appear on the screen of the CRTy. The greater the asymmetry in the electric field of the wave, the greater the difference in speed of the electron between hemisphere A and hemisphere B, the greater the deflection of the electron beam. The asymmetry in the electric field can be increased in three ways, 1) increasing the intensity while keeping the frequency of the wave constant, 2) increasing the frequency while keeping the intensity of the wave constant, and 3) increasing both. We need to increase the frequency while keeping the amplitude i.e., intensity of the electric field constant and check the effect. For this purpose, CRTx can be used to keep the intensity of the wave constant while changing the frequency as discussed earlier. By keeping the amplitude of the wave constant and increasing the frequency, the asymmetry of the electric field increases, so the deflection of the CRTy will increase and from this observation it can be determined that the magnetic force is increasing by keeping the intensity of the electromagnetic wave constant. Because according to prevailing electrodynamics the CRTy deflection is due to the magnetic field in that electromagnetic wave.

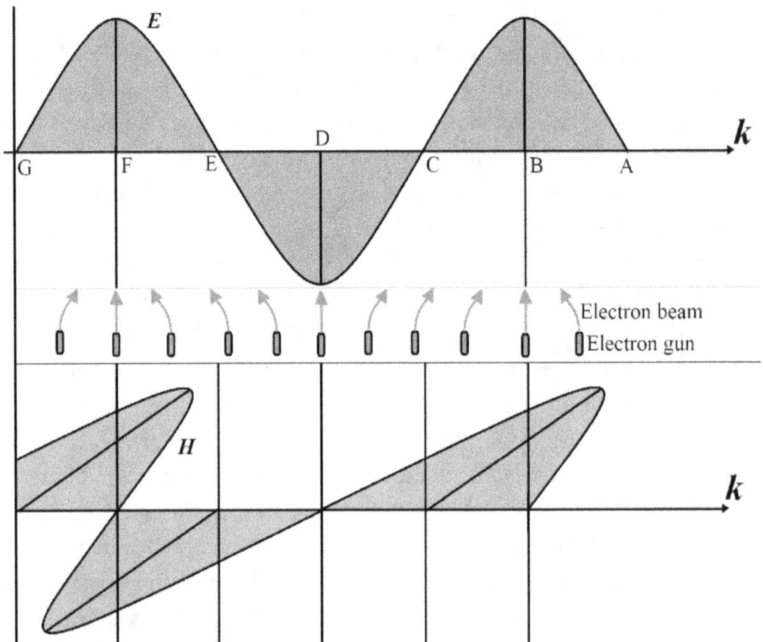

Figure 1.6.3. Possible magnetic field traced from the deflection of the electron beam of a CRTy at different sections of an electromagnetic wave.

Using this the following equation of magnetic force can be used to trace the phase-to-phase magnetic field in that electromagnetic wave.

$$\mathbf{F} = -e(\mathbf{v} \times \mathbf{B})$$

The magnetic force is in the direction in which the electron beam is deflected and the velocity of the electron is in the direction of the x axis. Using this the possible electric field and magnetic field, in phase to phase, of the electromagnetic wave are traced in Figure 1.6.3. Surprisingly, there is a 90-degree phase difference between the two. Because according to the Poynting vector in conventional electrodynamics, the phase difference between the electric field and the magnetic field is zero in vacuum as shown in Figure 1.6.5.(a). Of course, it is not mentioned anywhere that this has been verified by demonstration. And if the two are demonstrated to have a phase difference of 90 degrees, then the validity of Maxwell's equations, and indeed the existence of a magnetic field itself, can be questioned. To check this phase difference, a suitable single ramp signal can be applied to both CRTs as shown in Figure 1.6.5. This time the CRTx will trace the waveform of the electric field in the electromagnetic wave and the CRTy will trace the waveform of the magnetic field and directly verify how much phase difference there is between the two. Of course, the successful execution of this practical could lead to a change in human perception of magnetic fields, since magnetic monopoles do not exist in the universe, so magnetic fields must not exist independently.

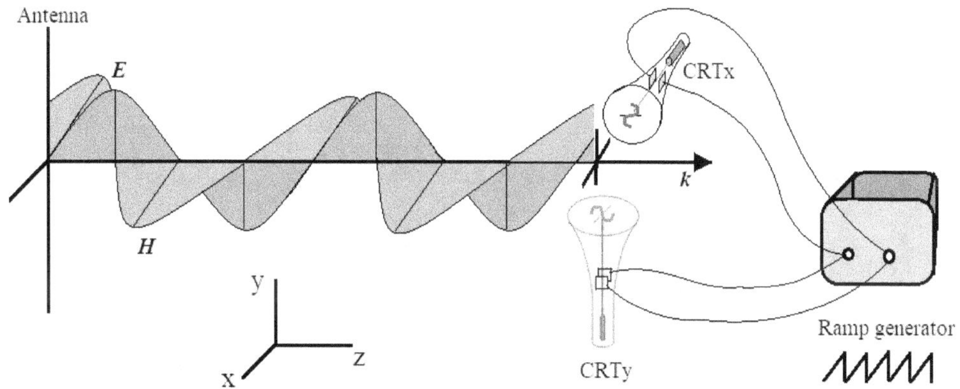

Figure 1.6.4. Experimental Design for Phase Investigation of Electric and Magnetic Fields in Electromagnetic Waves.

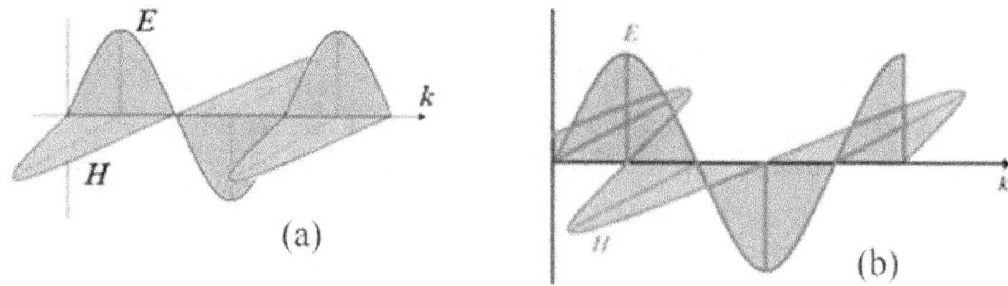

(a) According to the classical electrodynamics, electric and magnetic field waves are in phase.
(b) According to the explanation of Photoelectric effect, expected phase difference of 90 degree in electric and magnetic field waves.

Figure 1.6.5. Possible phase difference between the electric and magnetic fields in an electromagnetic wave, (a) according to prevailing electrodynamics, (b) according to the effect of asymmetric electric fields.

1.7 Demonstration of emission of photoelectrons in the plane of polarization of the light wave

From the discussion so far, we have seen that both the apparent force Fa which causes the electron to change its direction and the effective electric force F itself which accelerates the electron are in the plane of the electric field which is the plane of polarization of the light wave. That is, the electron emitted in the photoelectric effect should be found in the plane of the electric field of the light wave, i.e., in the plane of polarization. This effect can be seen when plane polarized light is used in the photoelectric effect. For this a special photoelectric tube has to be designed as shown in Figure 1.6.7, which has at least four anodes. The electronic circuit shown in the figure can be used to test this effect.

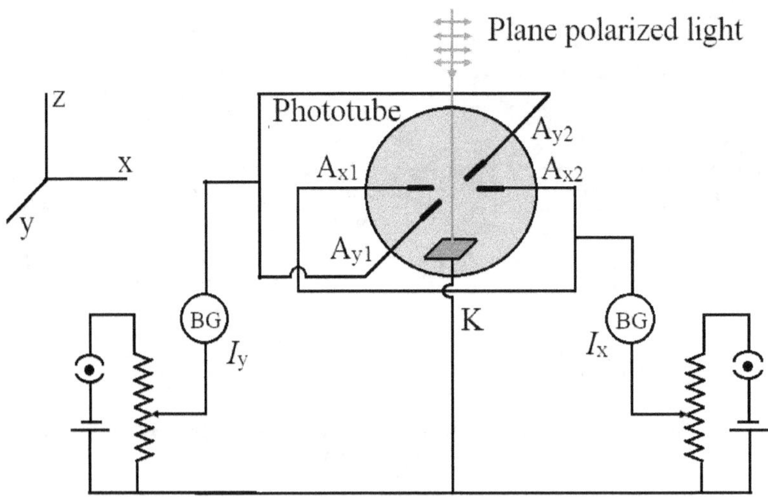

Figure 1.7.1. An experimental arrangement for investigating the emission of photoelectrons in the plane of polarization of light in the photoelectric effect.

When light is polarized along the x-axis, its electric field is parallel to the x-axis. If light falls on the cathode, the electrons will be emitted in the plane of the electric field. So naturally they will easily reach the anodes A_{x1} and A_{x2}, so relatively the ballistic galvanometer current I_x should appear more than the ballistic galvanometer current I_y. When the light remains polarized along the y-axis, the ballistic galvanometer current I_y should appear relatively larger than the ballistic galvanometer current I_x. From this it can be seen in which plane of light the electrons are emitting in the photoelectric effect. Special care has to be taken while preparing such a phototube and while conducting the experiment.

1.8 Summary

1. From 1887 to 1905 and even after that no one studied what could happen to the electron in its position in the photoelectric effect.

2. It was and still is accepted by all that light is an electromagnetic wave. Therefore, its energy must depend on the strength of the electric and magnetic fields. That is, the kinetic energy of the emitted electron in the photoelectric effect should depend on the intensity of the light. Therefore, increasing the intensity of the light should increase the kinetic energy of the emitted electrons in the photoelectric effect. But that doesn't seem to be happening. But as the frequency of the light wave increased, the kinetic energy of the electron increased, so it was understood that classical mechanics could not solve this problem. At that time everyone was thinking about the energy of the electromagnetic wave but no one seems to be thinking about how it is given or received i.e., what exactly happens there then and now. This is exactly the mistake made in the photoelectric effect.

3. No one disputes that force is mutual, but if an electromagnetic wave exerts a force on a charged particle, that particle must also exert an electromagnetic force on it. But there is no charge with that electromagnetic wave so the field of that particle must apply a force on the field in that electromagnetic wave. This means that the fields in the electromagnetic wave do not exert a force on the charge but apply a force to its field, so

the force must always exist through field-field interaction. But no one paid attention to this and the electric force could not be generalized back.

4. In any electromagnetic wave the electric field and magnetic field are asymmetric and it must affect the motion of the electron and no one has ever thought how it can. Of course, if we had tracked what could happen in the place of that electron, this thing would have come forward.

5. From the above discussion it is observed that electron follows curved path in asymmetric electric field. How curved that path is depends on how asymmetric the electric field is. This is what accounts for the photoelectric effect. The shorter the wavelength of the light or the higher the frequency, the greater the asymmetry of the field, the more likely the photoelectric effect is to occur. So it makes no sense to say that energy is concentrated at certain points in a light wave according to Einstein. This means that the light wave is a wave and also works in the photoelectric effect, thus raising the question about the validity of quantum mechanics.

6. This further suggests that the effect of an asymmetric electric field can be a magnetic field, meaning that the magnetic field does not have an independent existence. If this is true, then increasing the frequency of any electromagnetic wave must increase the magnetic force it applies. The success of this experiment will answer why magnetic monopoles are not found in the universe and reveal the true nature of magnetic fields.

7. Again this also shows that, even if the magnetic field is assumed to be the effect of an asymmetric electric field, there should be a 90-degree phase difference between the electric and magnetic fields in the electromagnetic wave which can be proven experimentally. This can trouble conventional electrodynamics, as this mechanics states that the electric field and the magnetic field in an electromagnetic wave are in phase. Of course, no one has experimentally tested this yet.

9. Equation for the photoelectric effect derived using an asymmetric electric field in light involve E_0, the amplitude of the electric field, requiring a thorough study of the shape of the light wave.

Chapter-02

Total length of single light wave emitted by sodium atom

2.1 Historical background

Light is such a gift of the universe that allows you to see the universe, experience the creation. This creates a clear image in our brain along with the surrounding color. We see such a world only and only because of illumination. These light rays i.e., electromagnetic waves are created from different atoms. We know that different colors have different frequencies and wavelengths, and we've calculated how much they are. All light rays travel at the same speed of 300,000 kilometers per second. It is believed that the speed of light is the fastest in the universe and no object can travel at or faster than this speed. This is called the law of spatial theory of relativity. In-depth research has been done on which light wave is emitted from which atoms. A light wave is a specific band of frequencies of electromagnetic waves that can be perceived by our eyes. When an electron in an atom jumps from its higher energy level to a lower energy level, the energy difference is emitted as an electromagnetic wave. Suppose that energy difference is E_3-E_2 then the frequency of that electromagnetic wave is $f = (E_3-E_2)/h$ where h is Planck's constant.

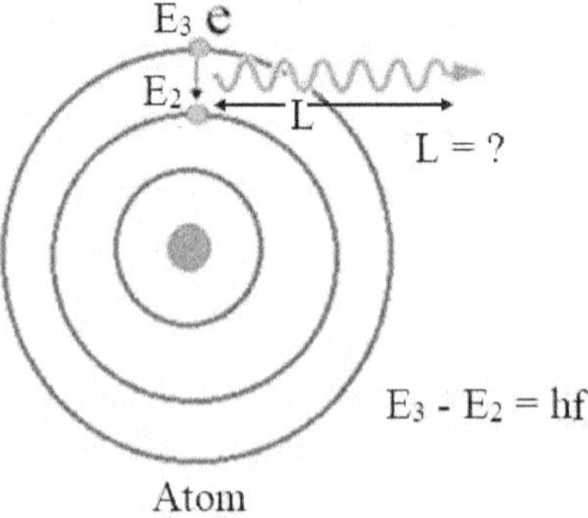

Figure 1.1.1. A single light wave resulting from the transition of an electron in an atom has a total length L and a frequency f.

If the frequency of that electromagnetic wave falls in the visible band, we can see that wave as what we call light rays. The funny thing is that despite all this research, nobody knows the total length of the light wave after it leaves the atom, nor is it documented anywhere, nor is anyone curious about it. The question that arises is whether our research is being conducted conveniently. It seems that research has come a long way and there is no research left in such generalities. But one thing is to be mentioned that no one knows how the energy difference is converted into electromagnetic wave after the transition of electron in atom and if this thing seems general and not important then it is better not to imagine your research mind. In fact, our research has gone in a typical direction and only the classical theory could have explained this but we have not been able to develop it in the right way. And now we are free to blame it on the impossibility as we are not in a position to answer it. There are many questions that are experienced in daily life or practice but we never notice which are fundamental and essential to understand the workings of the universe. Now we will discuss whether the total length of a single light wave can be measured.

2.2 Experiments in physics that give clues about the length of a light wave

Newton's ring experiment is used to determine the wavelength of a monochromatic light wave. It is commonly used to calculate the wavelength of sodium light in physics at the graduate level. Figure 2.2.1 (a) shows the structure of Newton's ring experiment and (b) shows the types of rings formed in.

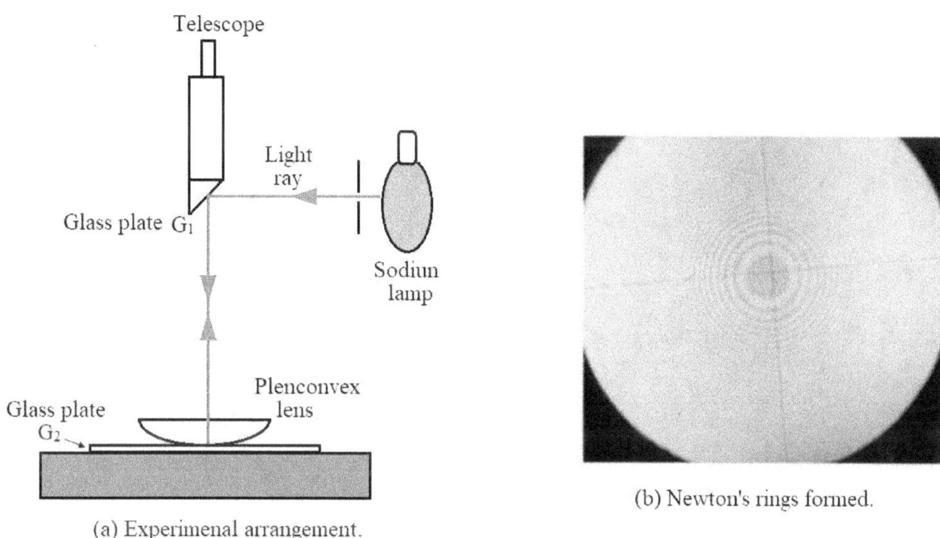

(a) Experimenal arrangement.

(b) Newton's rings formed.

Figure 1.2.1.(a) Structure of Newton's ring experiment and (b) the rings as seen through a telescope.
In this the wavelength of the monochromatic light is calculated by taking the diameter of the rings. The question is, what is the maximum number of rings that can be formed in this experiment? Many would say that any number can be created, since the lens used in the experiment is planoconvex. So the air gap between the lens and the glass plate underneath, which is responsible for the formation of rings, increases as we move away from the center of the lens, causing the rings to crowd from the center and eventually diffuse. So, they are not clearly visible and the total number of rings formed cannot be counted. Perhaps that is why the question of how many rings should be formed in total does not easily arise in our mind. In fact, in Newton's ring experiment only a limited number of rings form and their number must be equal to the number of crusts or troughs in a light wave emitted by sodium lamp. Amazing information is hidden in Newton Ring Experiment but surprisingly we all ignore it. This may also be because we are not clear about such things in our minds. In that we are very satisfied with the results obtained by solving the equations. But those equations don't necessarily solve everything that happens in that experiment, because they're purpose-driven, and they don't necessarily reveal all the secrets of the universe. Newton's ring experiment is designed to measure only the wavelength of monochromatic light. So obviously other information that can be obtained from it has been neglected. But a problem with the Newton ring experiment is that it is difficult to estimate the total number of rings formed because eventually the rings get crowded and faint near the edge of the lens. Another such experiment is the airways method, in which the diameter of a thin wire is measured using a monochromatic light source. As shown in Figure 2.2.2, it produces fringes but is not crowded. But here too, as we go from the first fringe to the higher numbered fringes, they become fainter and finally diffuse. So it is difficult to get

the exact number of the total number of fringes that have been formed, but it is possible to estimate the approximate number of fringes that have been formed. From this it can be estimated how many crests and troughs there are in a wave of monochromatic light used in a simple experiment and from this it can be estimated how long that wave will be. But no one seems to have thought much about it till now.

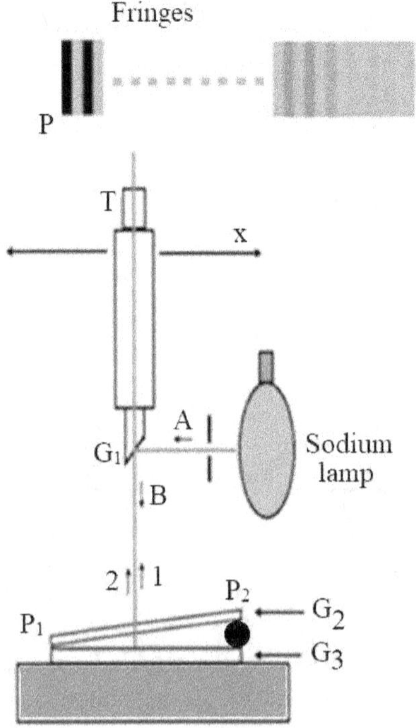

Figure 2.2.2. Experimental setup of Air-Wedge Method.

2.3 Air-Wedge method

Figure 2.3.1 shows the arrangement of airways method. In this, a ray of sodium light falls on a G_1 glass plate of the telescope which is transparent and is mounted at an angle of 45 degrees to the telescope. The light ray B reflected from it falls on two glass plates G_2 and G_3. Both these glass plates touch each other at one end and at the other end a piece of thin wire is placed between them whose diameter is to be drawn. In the two glass plates G_2 and G_3, the air included from point P one to point P two increases and is wedge-shaped. So, this experiment is called the Air-Wedge method and this is the reason for the formation of fringes. In this, the fringes are formed due to the path difference of light ray 1 reflected from the lower surface of the G_2 glass plate and light ray 2 reflected from the upper surface of the G3 glass plate. As we move from point P_1 to point P_2, the path difference between them increases and they interfere successively constructive and destructive producing successively dark and bright fringes. When the path difference between the two becomes so large that they become independent and cannot interfere, fringes cannot form. The spectrum in the subsequent part will be diffused. If we want to understand the process of formation of fringes in detail, a single light wave generated from one atom of sodium lamp at a time has to be used for fringe formation. Let us assume that as wave A which is the first to fall on a glass plate G and its reflected part is B. Now A wave

24

and B wave are going to have the same length and wavelength. Some part of wave A part is transmitted through the glass plate and remaining part is reflected. The reflected part is Wave B. Then there is no clear mention of what exactly should be the difference between Wave A and Wave B. At least there should be a difference in amplitude between the two, and if so, it should result in the photoelectric effect. But this effect is applicable for maximum kinetic energy of the photoelectrons that means the fact that we are still ignorant about the exact shape or structure of the light wave. But there is no record anywhere of investigating the photoelectric effect by using a single light wave with varying its amplitude. But a question arises that there is a single light wave emitted by an atom at a time i.e., it corresponds to a single photon with energy hf. Now that wave is divided into two parts, one transmitted which has frequency f means that it also corresponds to a photon with energy back hf and the other reflected means the reflected wave b which has frequency f and also corresponds to a photon with energy back hf. So, two photons are created from one photon? In fact, the common man should not ask such questions because he does not know quantum mechanics. Physicists would say, have you properly understood de Broglie's hypothesis of matter wave? Have you read the Heisenberg Uncertainty Principle? Did you solve the Schrödinger equation? This means that the only option is to understand them enough to accept them. What you think doesn't matter anymore now but matters the quantum mechanics. Does the universe behave differently at different levels? Never. We did not understand him properly. And this has to be understood sometime. Let us return once again to the fringes produced by a single light wave.

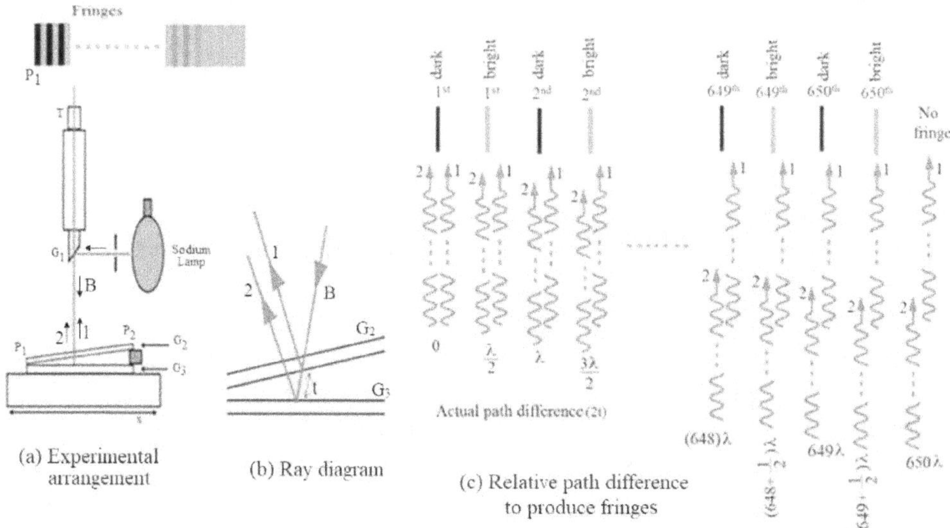

Figure 2.3.1. Ray-diagram and path difference formation of interference fringes in Air-Wedge method.

Then the reflected ray or wave B will fall on the glass plate G_2 and some part will be reflected from its upper surface and some part will be transmitted below. A portion of the transmitted wave will be reflected back from the lower surface of the glass plate G_2 will be wave 1 one and the wave transmitted back down and reflected from the top surface of glass plate G_3 will be wave 2. Depending on the path difference between wave 1 and wave 2, those waves will interfere constructively or destructively and create fringes. To test the maximum number of fringes produced in this system either the telescope has to be moved from point P_1 to point P_2 or the base, on which the glass plate is mounted, can be moved left and right without moving the telescope. The second option will be suitable so that light source does not need to be adjusted each time. When wave B hits the glass plate at point P, wave 1 will reflect from the bottom surface of glass plate G_2, adding a 180-degree phase difference to the wave because it is reflecting through a medium with a higher refractive index. The wave transmitted from glass plate G_2 will be reflected from its upper surface at glass plate G_3, which is wave 2. Wave 2 will not add any phase difference because it is reflecting through a medium of low refractive index. Since the glass plates G2 and G3 touches at point P_1, the path difference will be 2t = 0 and the total path difference will be $(2t + \lambda/2) = (0 + \lambda/2) = \lambda/2$. Due to this, both the waves will be 180 degrees out of phase with each other, thus interfering destructively and creating a dark spot. The reason it is called spot is that we are assuming that only two waves are interfering. In an actual experiment the number of light waves involved will produce fringes as shown in Figure 2.3.1. Now if the base continues to shift to the left, the path difference between the two waves will increase and when it becomes $2t = \lambda/2$, the total path difference will be $(2t + \lambda/2) = (\lambda/2 + \lambda/2) = \lambda$. The two waves will then be in phase and constructively interfere, creating the first bright spot. Moving the base further to the left will again increase the path difference between the two waves and when it becomes $2t = \lambda$ then the total path difference will be $(2t + \lambda/2) = (\lambda + \lambda/2) = 3\lambda/2$, then again, the two waves are out of phase, they interfere destructively and another dark spot is formed. Going further the path difference will increase when it becomes $2t = 3\lambda/2$ so the total or effective path difference will be $(2t + \lambda/2) = (3\lambda/2 + \lambda/2) = 2\lambda$ then the two waves will be in phase and will constructively interfere thus forming another bright spot. As we move away from the point P_1, dark and bright spots will continue to form. One thing to note here is that as we move away from the P_1 point, the overlapping portion of both the waves that cause the dark and bright parts to form will decrease so the dark and bright spots will fade. Suppose wave one or wave two has a total of n troughs and n crests, then the last n dark spot and the subsequent n bright spot will form, and then the two waves will become independent. Those two waves will not interfere in this next region and no bright or dark spot will be created. The next section will have a diffused color of sodium light. One thing is clear from this that if a single light wave has a total of n troughs and n crests, then n number of dark and n number of bright spots are formed. From this we can tell how many troughs or crests there are in that wave and what is the total length of that wave. Our objective is to determine the total length of a wave emitted by an atom in a sodium lamp. A sodium lamp contains a large number of atoms and all of them emit waves at the same time, so different combinations of emitted waves can be produced. Also, sodium lamp will produce dark and bright fringes in the interference pattern as shown in figure 2.3.1 instead of dark and bright spots. They cannot be precisely measured but when a sodium lamp is used approximately six hundred to seven hundred appear, so an average of 650 can be assumed. A large number of simultaneous waves are emitted from sodium lamp and can join each other in their journey, increasing their total length. To examine the difference in their fringe formation, for example, we consider two waves W_1 and W_{12} having n number of troughs and n number of crests as shown in figure 2.3.2. If both waves join in phase, W' wave will form and if they join out of phase a W" wave will form. The probability of formation of both the waves is going to be the same. Therefore, both types of waves will be equally involved in creating the interference pattern. No doubt, they will form the final nth dark and nth bright fringes as discussed above. Then, as shown in Figure 2.3.3, at the next $\lambda/2$ path difference, W' will produce a dark fringe while at the same time W" will produce a bright fringe. That means the total diffuse color will be. Then at the next $\lambda/2$ path difference, as shown in Figure 2.3.4, W' will produce a bright fringe while at the same time W" will produce a dark fringe. So again, overall, there will be diffuse color. That means the next region will have a completely diffuse color. This means that if the original waves emitted by atoms in a monochromatic lamp have n crests or n troughs or n wavelengths, the resulting fringe pattern will have only n dark and n bright fringes. No matter how many waves emitted by the atoms in the lamp join each other during travel and increase in length, they will have no effect on the total number of fringes produced.

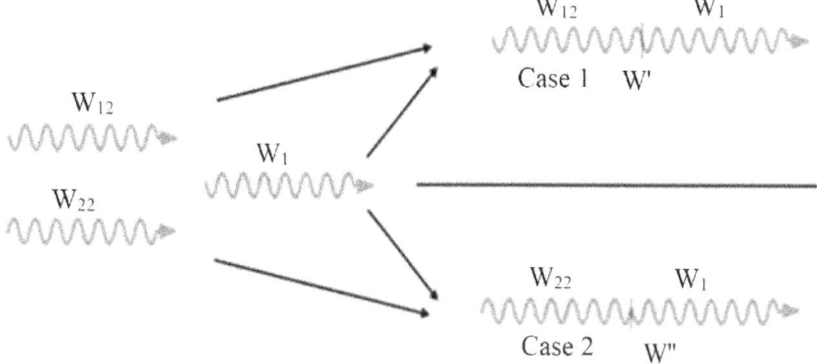

Figure 2.3.2. Waves formed by two waves joining together.

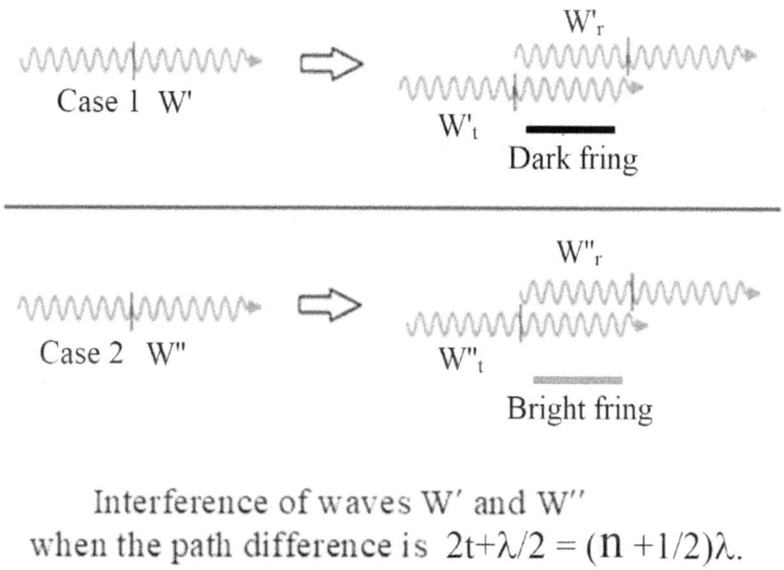

Interference of waves W' and W"
when the path difference is $2t+\lambda/2 = (n +1/2)\lambda$.

Figure 2.3.3. The fringes produced by W' and W" when the waves have $2t = n\lambda$ path difference.

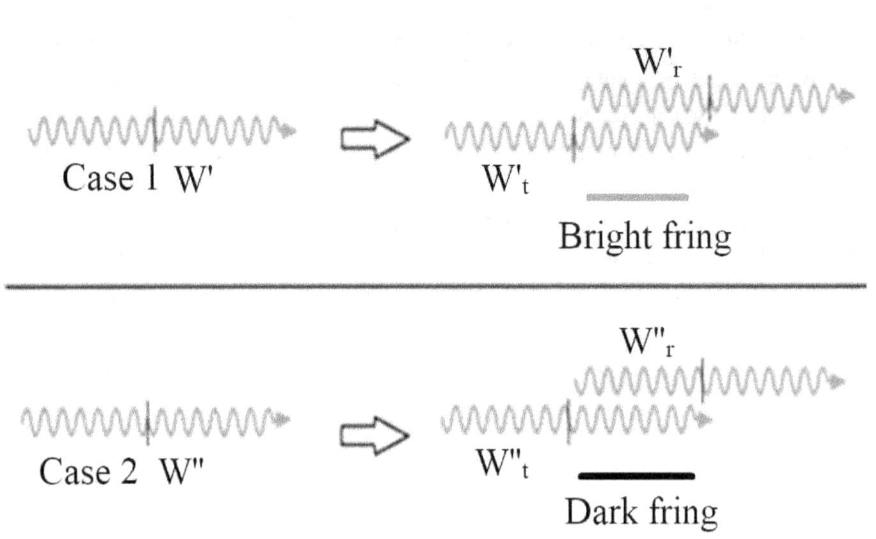

Interference of waves W' and W" when the path difference is $2t + \lambda/2 = ((n+1/2)+1/2)\lambda$.

Figure 2.3.4. Fringes produced when W' and W'' have $2t = (n+1/2)\lambda$ path difference.

When this experiment was conducted using a sodium lamp, approximately 650 bright fringes were observed in the interference pattern. The thickness of the object placed in the two glass plates has to be adjusted to make the fringes appear neat and clear. Of course, the fringes diffuse and disappear as you move away from the P_1 point, making accurate fringe measurements difficult. For this, by counting how many fringes there are in one centimeter, the total number of fringes can be estimated by estimating how many centimeters are formed. Approximately 650 dark and bright fringes are observed using sodium light. Sodium light has two wavelengths i.e., $\lambda_1 = 589$ nm and $\lambda_2 = 589.6$ nm and the airways method cannot distinguish between them due to the very short difference between them. So, in this method only one wavelength can be assumed which is mean wavelength $\lambda = 589.3$ nm. So, the total length of light emitted at one time by one atom in sodium should be

$$L = n \times \lambda = 650 \times 589.3 \text{ nm} = 383045 \text{ nm} = 0.383 \text{ mm}$$

This length is much longer than expected. One wonders how such a long light wave can be expressed as a photon, which is considered as a point particle. In the photoelectric effect, that photon absorbs an electron. The size of an electron is considered to be less than 1 nanometers and the wavelength accompanying that photon is 383045 nanometers, and if an electron absorbs that photon, it must absorb the entire wave. It is unbelievable how such a long wave can be absorbed by an electron whose size is less than 1 nanometers. So, it is impossible for a light wave to manifest as a photon. Another thing is that the light wave is an electromagnetic wave and it has 650 crests or troughs, so the transition of the electron in the sodium atom that created this wave must vibrate at least 650 times in one place. Otherwise, such a wave cannot be formed as anyone knows. A model of the atom that has been put forward to society through various popular means is the Rutherford-Bohr planetary model in which, as shown in Figure 2.3.5 (a), the electrons move in orbits around the nucleus. Obviously, this is the perception in everyone's mind. If this is the case, then electrons cannot produce an electromagnetic wave of such length. Given that quantum mechanics understanding, electrons can be expressed as waves, as shown in Figure 2.3.5 (b), how can one wave, which is a scalar, produce another electromagnetic wave? So, this gives a clear indication that electrons in an atom should not move in orbits or even manifest as waves. They must be stable regardless of what happens in the atom. This probably uses the spin velocity of the electron which we will see later in the spin atomic model. A drawback in this experiment is that the exact length of a light wave emitted by an atom at a time cannot be measured, only an estimate can be made, but it also provides a lot of information. In this, it is necessary to conduct a comparative study by measuring the total length of the wave emitted by the atoms of matter in solid, liquid and gas form. It is also necessary to study whether the temperature of the material has any effect on the total length of the wave. This can reveal many things that we have never thought of and can give a new direction to research.

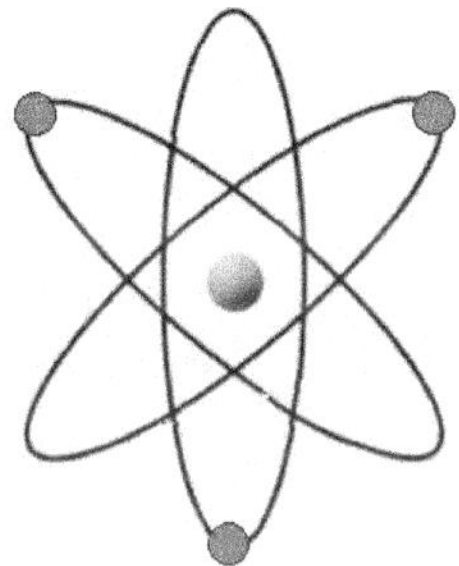

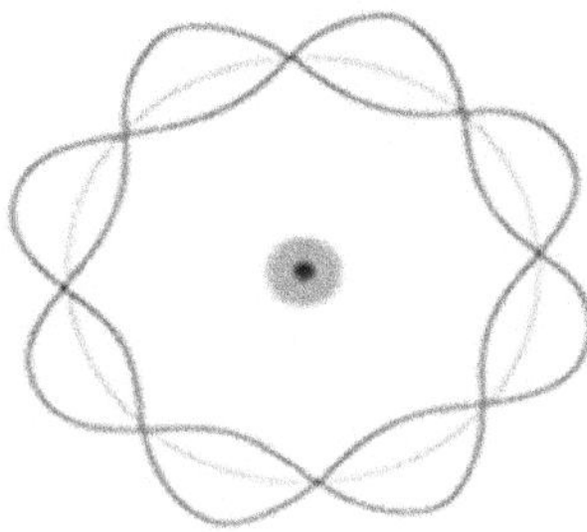

(a) Rutherford-Bohr
model of atom.

(b) Quantum mechanical
model of atom.

Figure 2.3.5. Models of the atom, (a) Rutherford-Bohr planetary model, (b) quantum mechanical wave model.

2.4 Partial reflection and partial transmission of light rays

Both interferometers and diffractometers are methods that prove that light has a wave nature. But no one seems to have done much research or in-depth study of what the shape of the light beam emitted from one atom at a time might be. In an interferometer, two light waves are produced from a single light wave and brought together at a single point by creating a path difference between them. Bright or dark spot is produced depending on the path difference between them and whether they are in-phase or out-of-phase with each other. It uses a concept called partial reflection and partial transmission when one light ray is formed into two light rays. In interferometers there is partial reflection, meaning it is not clear what exactly is happening. That is, it is avoided to explicitly mention whether one light wave produces two light waves. In some places it is mentioned that some part of light ray is partially reflected and, in some places, it is said that some part of every light ray is partially reflected. There is a big difference between the two sentences. The first sentence means that there are many rays, some part of them is reflected, and the second sentence says that some part of each ray is reflected. The fringes are produced only when the second sentence occurs in the interferometer. But many avoid mentioning this clearly. Because the production of two light rays of one light beam means the production of two photons of one photon means doubling of energy. This discrepancy is Einstein's distrust of light quanta and our inability to help ourselves. So sometimes we have to understand photoelectric effect properly. At least one thing we learned from this is that as the frequency of an electromagnetic wave increases, the magnetic force applied by it increases. Now it is necessary to understand what the exact structure i.e., shape of the light ray can be. One thing observed from interferometer is that one light beam can become two light beams, which have the same frequency and the reason for this is partial reflection and partial transmission. And the expected fringes are formed only if the interferometer bisects each light ray and participates to form the fringes otherwise not. If we consider the Air-Wedge method, a light beam I from a sodium lamp, which is a single light wave emitted at one time from an atom, will first partially reflect from the glass plate G_1, as shown in Fig. 2.4.1, will have part I_2 and transmitted part I_1. A part of the light ray I_2 will be further reflected from the upper surface of the glass plate G_2 as I_3 and a part of the transmitted ray will be reflected back from the lower surface of the glass plate G_2 as I_4 and a part of the transmitted ray will be reflected back from the upper surface of the glass plate G_3 will be I_5 and the remaining part transmitted will be I6. This means

$I = I_1 + I_2 + I_3 + I_4 + I_5 + I_6$

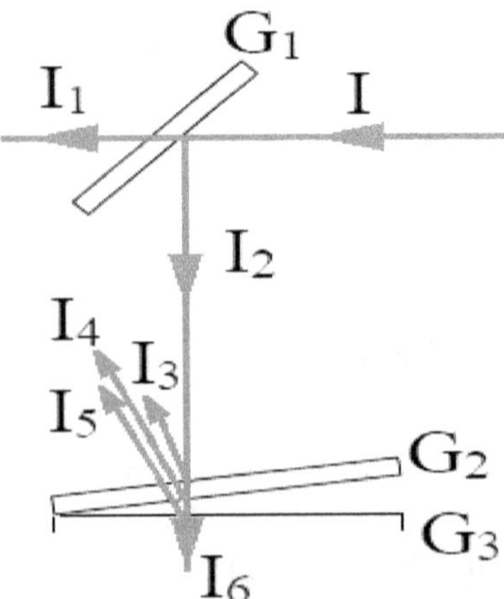

Figure 2.4.1. Partial reflections and partial transmissions of a light ray.

If a ray of light can be divided so much, surely its intensity must decrease per division. Here only two rays I_4 and I_5 are involved in forming fringes and the fringes formed will be visible only if the intensity of both these light rays are at the same level. Suppose we make I6 zero by using a mirror instead of a transparent glass plate G_3, then the intensity of I_5 will increase more than I_4, so the fringes will not be formed properly. So that instead of a mirror, a glass plate is used and the ray transmitting out from it is absorbed by placing carbon paper under it. Now a light ray is an electromagnetic wave, which is produced by the transition of an electron in an atom. Although it is not clear how exactly it is produced, it is a universal truth that the corresponding electron must oscillate in a position for an electromagnetic wave to be produced. The frequency at which it oscillates will produce an electromagnetic wave, and the electron's electric field is essential to produce it, which is constant. When an electron oscillates, the shape or amplitude of the wave produced by it will depend on the maximum distance it travels from its mean position. This involves understanding what exactly the intensity of a light wave can be. We know that the transition of an electron in an atom creates a light ray and is an electromagnetic wave. However, as we have seen in the photoelectric effect, a magnetic field can be an effect of an asymmetric electric field, so a light wave must have only one field, and that is the electric field. Another thing is that in order for an electric field wave to be formed, the corresponding electron must oscillate at that frequency and while oscillating, its kinetic energy and potential energy must be converted into each other. Also, disturbances in the electric field created by those oscillations must be propagated in the form of waves. Naturally, as disturbances in the electric field propagate as waves, the total energy of the oscillating electron must decrease and eventually stabilize. Figure 2.4.2 (a) shows the time-dependent oscillations of the electron while Figure 2.4.2 (b) shows the resulting electric field wave.

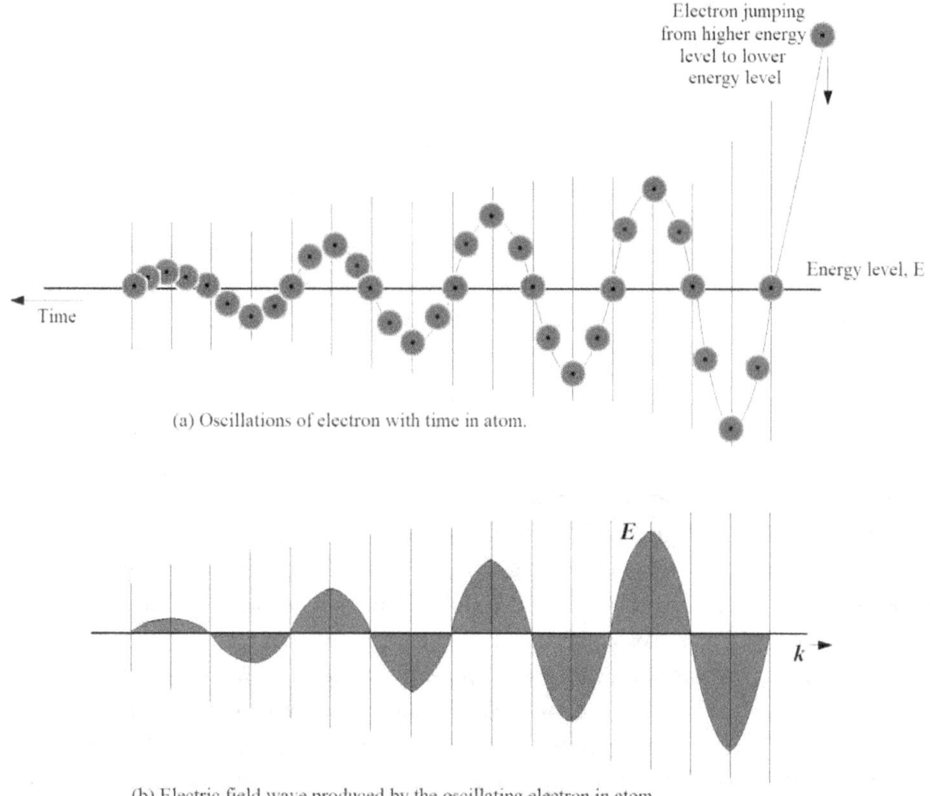

Figure 2.4.2 (a) Time-dependent oscillations of electrons in an atom, (b) Electric field wave produced by oscillations of electrons in an atom.

Here the frequency of oscillations will be constant but the amplitude should decrease. But we have seen that a light wave emitted by a sodium atom has about 650 crests and 650 troughs. That means the amplitude of the wave will decrease slowly. To create such a wave requires a medium which we have already called 'Ether'. We will discuss this in detail in the chapter 'Ether, Fundamental Particles and Fields'. When a light beam i.e., electric field wave is split due to partial reflection and partial transmission, their amplitude should be divided into two parts then its effect should be seen in photoelectric effect. But since this effect is checked for the maximum kinetic energy of the photoelectron, it might have been checked for the maximum amplitude of the electric field wave. But if the shape of that electric field wave is in the form of a tube, like the electric field waves in a laser beam, it remains constant no matter how far it travels in space and until it does not interact with any charged particles. In turn, the field of such an electric field wave will depend on how far the corresponding electron, which was creating the field wave, was displaced from its mean position at that time. That is what we call amplitude. Also, the width of that wave should be equal to the size of the effective electric field of the electron. So, the shape of a light wave, i.e., the size of its electric field, will depend on the conditions under which it is produced which is important to investigate. Again, for such an electric field wave to form, the corresponding electron must oscillate in a position such that it cannot move in an orbit within the atom. Then how it can use its spin motion to settle in an atom at a certain distance from the nucleus is discussed in detail in the next chapter, 'Spin Atomic Model'.

2.5 Summary

1. It turns out that light waves can explain the photoelectric effect as electromagnetic waves. So, it is necessary to know the shape of single light wave. If Einstein says that a light wave is a photon and it absorbs an electron, then the length of that wave must be known.

2. A light wave is produced by the transaction of electron in an atom. There was not a wave before the transaction of the electron and not after the transaction was completed. This means that the wave was generated over a period of time, so it is clear that the wave will have a finite length.

3. Newton's ring experiment to find the frequencies of monochromatic light is done in all in physics laboratories in the world of physics at UG level. It is surprising that no one asks the simple question of how many rings should be formed, the answers of which contain very important information. The number rings in Newton's ring experiment are expected to be equal to the number of crests or troughs in the monochromatic wave used from which the length of that light wave can be estimated.

4. A light wave emitted by a sodium atom usually consists of 650 crests and 650 troughs, so that the total length of the wave becomes 383045 nm. So, the question arises as to how an electron whose size is less than 1 nm can absorb such a wavelength. Therefore, there must be an interaction between the fields in this wave and the electric field of the electron, which can be called field-field interaction.

5. For a wave of such length that has 650 crests, the corresponding electron in the sodium atom must vibrate 650 times in one place while producing the wave. This means that the electrons in an atom should not move in orbits. According to quantum mechanics, if the electrons in the atom are expressed as waves, they will not be able to produce light waves of such a length. That is, in order for electrons to be stable in an atom, it is necessary to think differently about how they can remain stable.

6. To understand the photoelectric effect it is necessary to know the shape of the light wave with which the electrons interact. Hence, it is necessary to conduct research from this point of view which is left out.

Chapter - 03

Spin Atomic Model

3.1 Historical background of atomic models

The idea that matter is made up of different units is very old, dating back to ancient cultures including Greece and India. The word atom, meaning un-divisible, was first used by Socrates (470–399 BC), Democratis (460–370 BC) and his mentor Leucippus. Democratis used to say that atoms are infinite in number and are eternal. The properties of each substance depend on the atoms of which it is composed. Also, in ancient Indian culture Kanada (obscure, 6-2 BC) was an ancient Indian naturalist and philosopher who founded the 'Vaisheshika school' of Indian philosophy which represents the earliest Indian physics. Physics was central to Kanad's assertion that everything perceptible is based on motion. This underlines the importance he gave to physics in the understanding of the universe. He says that the atom must be spherical because it must be equal in all dimensions. He postulated that all matter is composed of four types of atoms, two of which have mass and two are massless. He is known for developing the foundations of an atomistic approach to physics and philosophy in his Sanskrit treatise 'Vaisheshika Sutra'. In the subsequent period, the atomic theme is nowhere to be found commented upon, even until the twelfth century. In the twelfth century, while studying some old documents, some of Aristotle's statements about matter came to light, namely that matter is continuous and infinite and can be subdivided without limit. Further, it was not until the 19th century that anyone could draw any solid conclusions about the atom. But it was generally accepted that the atom is the smallest unit of matter and cannot be further subdivided.

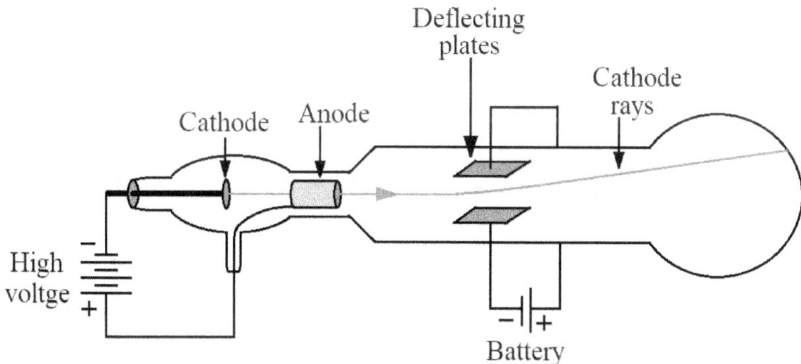

Figure 3.1.1. Crook tube used by JJ Thomson to produce cathode rays.

But in 1897, JJ Thomson generated a cathode ray by applying a high voltage across the electrodes in a vacuum tube, called a Crooke tube, as shown in Figure 3.1.1, and a glow patch was formed after it fell on the front screen. They could be deflected towards the positively charged plate as shown in the figure. From this it was concluded that they must be negatively charged corpuscles, henceforth called electrons. Georg Stonia discovered their mass to charge ratio which was found to be 1800 times smaller than that of a hydrogen atom. Thus, for the first time, cathode rays, i.e., electrons, were discovered. This led Thomson to suggest that atoms can split and that these corpuscles, i.e., electrons, must serve as the building blocks of the liquid. From this he argued that the atom, which is neutral, must have electrons scattered in a sea of uniform positive charge, leading to the so-called plum pudding model, which is shown in Figure 3.1.2.

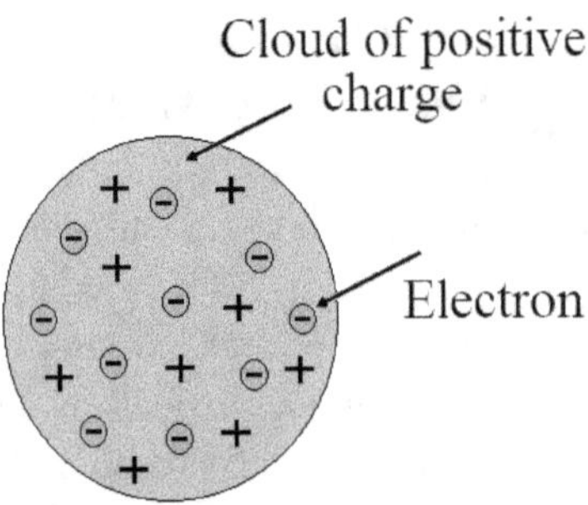

Figure 3.1.2. Plum Pudding Model.

Ernest Rutherford was a New Zealand physicist who studied at Cambridge University under Thomson. It was his later work at the University of Manchester that provided greater insight into the interior of the atom. He devised an experiment, Figure 3.1.3. (a), to investigate the structure of the atom in which positively charged alpha particles were incident on a thin sheet of gold and the scattered beam was investigated. It found that the positive charge and most of the mass of an atom is in the centre, and the lightly negatively charged particles are distributed around the centre. Therefore, he suggested a planetary model, Figure 3.1.3. (b), proposed in which electrons revolve around a small, positively charged nucleus.

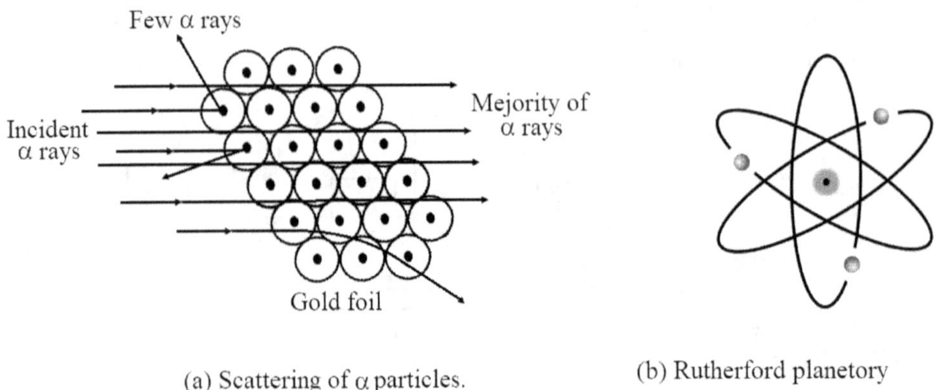

(a) Scattering of α particles.

(b) Rutherford planetory model of atom.

Figure 3.1.3. (a) Rutherford experiment of alpha particles, (b) Rutherford planetary model of atom.

Rutherford was on the right track, but his model could not explain the emission and absorption spectra of atoms and why electrons do not fall into the nucleus. In 1913, Niels Bohr proposed the Bohr model, which states that electrons only orbit within a certain distance from the nucleus. According to his model, electrons cannot fall straight into the nucleus but can make quantum jumps between energy levels. He postulated the existence of stationary orbits. Electrons can only be found in these specific energy levels. In other words, their energy was quantized, and couldn't take any arbitrary value. Electrons can jump between these energy levels, called 'stable states', but do so by absorbing or emitting energy. That is, he said that the electron in the hydrogen atom moves in certain orbits called stationary orbits in which the electron does not radiate energy while moving. Also, it can jump from one stationary orbit to another stationary orbit. If it jumps from an outer orbit to an inner orbit, it will radiate energy and if it jumps from an inner orbit to an outer orbit, it will absorb energy. That is, it cannot move into any other orbit except the stationary orbit, as if it were the mind of the electron.

But Bohr's model did not solve all the problems of the atomic model. This model proved useful for hydrogen atoms, but did not work for larger atoms. It also violates the Heisenberg Uncertainty Principle. Thus, Bohr's model still needed to be refined. But the concept of the Rutherford-Bohr planetary model of the atom based on the motion of the planets as shown in Figure 3.1.4 took root in the public mind.

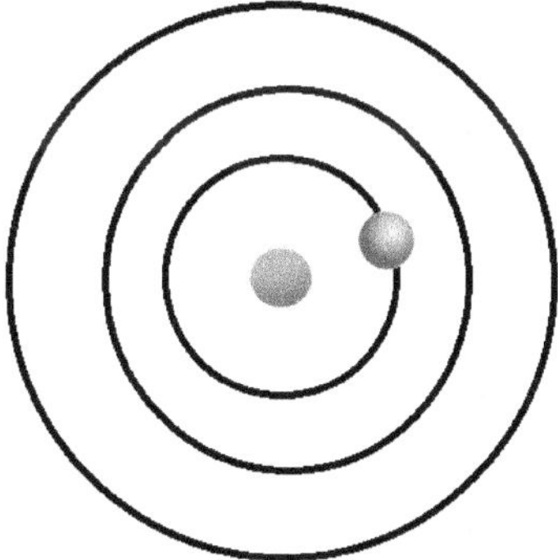

Figure 3.1.4. Niels Bohr Model of Hydrogen Atom.

At this point, many scientists were investigating and trying to develop a quantum model of the atom. Solving the problem of the photoelectric effect, Einstein published his paper in 1905 that provided a revolutionary explanation for the photoelectric effect. This was based on his "light quanta" hypothesis. He proposed that light behaves as a stream of discrete, localized units of energy called 'light quanta'. Each light quanta have an energy hf where f is the frequency of the light wave and h is Planck's constant. Just as light waves can behave like particles i.e., light quanta, a particle must behave like a wave. This proposition was first proposed by L. de Broglie in the form of 'the dual nature of Matter'. That concept was further used by E. Schrödinger in 1926 to propose that instead of electrons moving in stationary orbits in an atom, they can behave as waves, as shown in Figure 3.1.5.

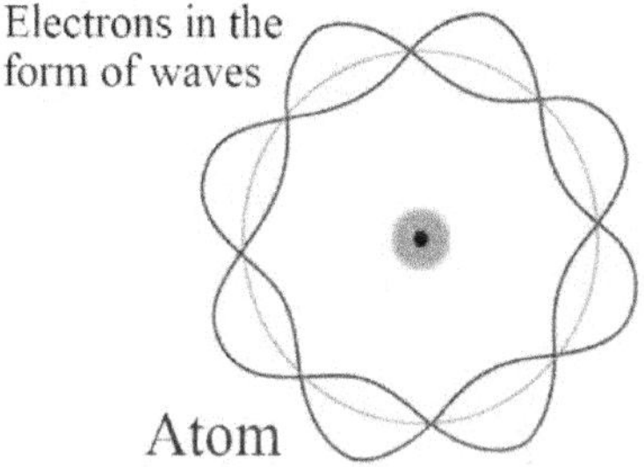

Figure 3.1.5. Electrons in atoms can behave as waves.

Schrödinger solved a series of mathematical equations to model the distribution of electrons in an atom. His model shows the nucleus surrounded by a clouds of electron density. These clouds are where electrons are most likely to be found, although we don't know exactly where they are. These regions in space are called electron orbitals. This atomic model of Schrödinger is called cloud model of atom.

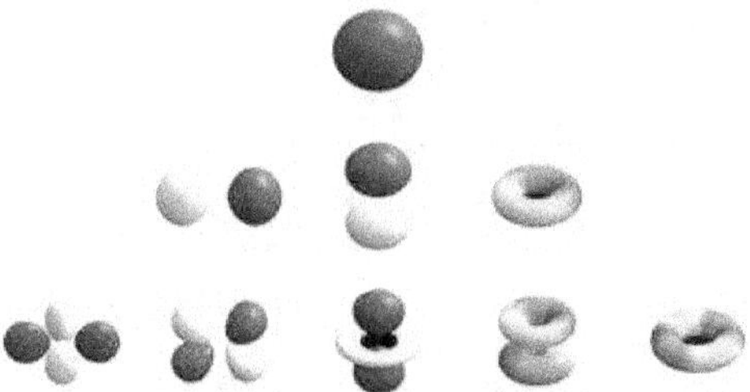

Figure 3.1.6. Orbitals in atoms that indicate the probability of finding electrons.

3.2 Inconsistencies in the Prevailing Atomic Model

No one can say with certainty which atomistic model is widely accepted in the current situation. One such model seen in use to some extent through various mediums and logos of many atomic energy related organizations is the Rutherford planetary atomic model shown in Figure 3.1.3. (b), in which the electrons are expected to move in circular orbits around the nucleus. Another atomic model that seems to be gaining some acceptance in the scientific world is that electrons exist in wave form, as shown in Figure 3.1.5 It is also known as cloud model of atom as the location of electrons is represented in the form of orbitals as shown in figure 3.1.6. That means electrons are in motion in both models. In the first model, electrons are assumed to move in orbits and in the second model, electrons are assumed to behave as waves, which means that they must be in motion according to de Broglie's hypothesis. Either one or a combination of these atomic models will serve as the building blocks of our matter. The first question that should arise in the common man's mind is that most substances are solid in shape and do not change easily, so if all the electrons in the atoms are moving in orbits or if all those electrons are expressed as waves, then how can such atoms form solid matter as shown in Figure 3.2.1?

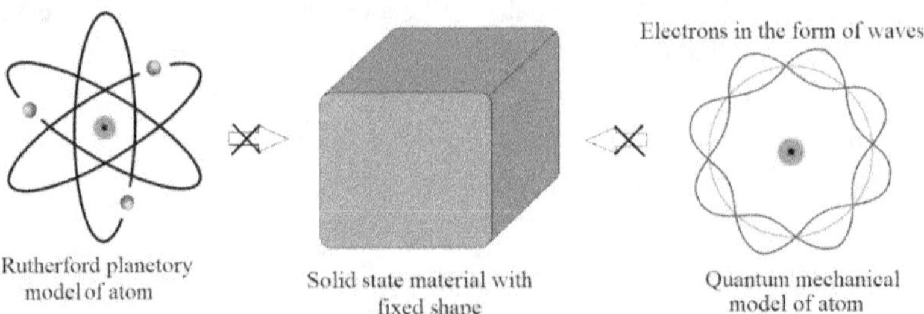

3.2.1. Rutherford's Planetary Model of Atom and Quantum Mechanical Model of Atom Impossible to form solid matter.

No matter how much we think about it, this is not possible. No matter how hard we try, this is not possible. Even a common man would clearly say the same. Here the common man is mentioned because his mind will not be encumbered by anything, he will think straight from his emotions. But the people who work in this field, their intellects are heavily encumbered by prevailing knowledge, so they lose the ability to think freely and keep moving in a certain direction. And because such people dominate this field, many things can not be deduced rationally. The fact is that solid-shaped matter cannot be formed from either of the above-discussed atomic models. Another thing worth mentioning is the light wave that originates from the atom. For example, we have seen that a sodium atom emits light waves with about 650 crests or troughs at a time. If the electron emits a light wave with so many crests or troughs, after the electron jumps from a high energy level to a low energy level in that atom, it is expected to oscillate in the same place according to the number of crests. If an electron moves in an orbit in an atom, it cannot stop at one place and oscillate. Also, if it is present in the atom in the form of a wave, it is not possible to create another wave from it. So, both these atomic models are not in a position to support the generation of light waves i.e., electromagnetic waves. Another thing that needs to be discussed is that the net magnetic moment of a paired electron in an atom is zero. Preferably an atom has two electrons in pairs so that their net magnetic moment is zero. At present two motions are assumed for electrons in an atom namely orbital motion and spin motion. First considering the orbital motion it revolves around the nucleus in a circular motion as shown in figure 3.2.2 (a). It is expected that the centripetal force and centrifugal force of the electron will balance and the electron will continue to revolve around the nucleus in a fixed orbit. Another strange thing to mention here is that if the electron continues to orbit, it will radiate energy so that its own energy will gradually decrease and finally it will fall into the nucleus. But a solution to this proposed was that if it moves in a certain orbit, called stationary orbit, it will not radiate energy and it will not fall into the nucleus. It will continue to move in the same orbit. Now what is on the electron's mind that if it moves in this orbit, it does not have to radiate energy and if it moves in that orbit it radiates energy. To determine whether an electron has a mind or emotions, in which orbit it must radiate energy. If there is no way out, the matter should be left open for discussion, but wise men are wise because they are silent at the right time, and when they are over-imaginative, the world seems different and they keep talking passionately. If an electron moves in an orbit, it must move around the center of the nucleus for sure, otherwise its centripetal force and centrifugal force will not balance. If an electron is moving in an orbit, its orbital magnetic moment would be,

$$\mu = -\frac{e\,r\,v}{2}$$

where r is the radius of the orbit and v is the velocity of the electron in the orbit and its direction is as shown in Fig. 3.2.2 (a). If two electrons are in a pair, they must have the same energy, i.e., they must move in the same orbit. For their effective magnetic moment to be zero, they must move with the same speed but in opposite directions and in the same orbit as shown in Figure 3.2.2 (b) if and only then their effective orbital magnetic moment will become zero.

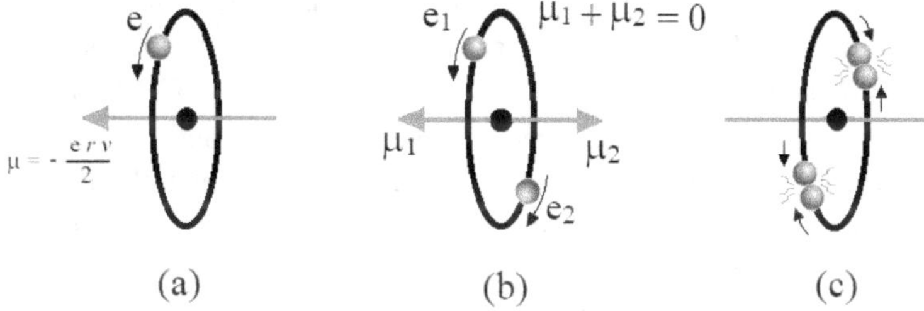

Figure 3.2.2. (a) Orbital magnetic moment of an electron, (b) When two electrons are moving in the same orbit in opposite directions with the same speed, their net orbital magnetic moment is zero, (c) When two electrons are moving in the same orbit in opposite directions with the same speed, they will continue to collide.

So, it moves in the orbit and will come close twice while completing one revolution at a time. Since they both have repulsive forces, they will repel each other, so under no circumstances can two electrons move in the same orbit in opposite directions. But the net orbital magnetic moment of the paired electrons is zero, so the question arises as to whether they actually move in orbital motion. Because the concept of orbital motion was derived from planetary motion to explain how an electron can remain at a certain distance from the nucleus, in which the centrifugal force and the centripetal force balance. If another option allows the electron to remain stationary at a certain distance from the nucleus, then the option of orbital motion for the electron may be nullified.

$$\mu = -\frac{e}{2m} L$$

Electron in spin mtion

(a)

$$\mu_1 + \mu_2 = 0$$

$\mu_2 \quad \mu_1$

e_2 $\qquad e_1$

(b)

3.2.3. (a) Spin magnetic moment of an electron, (b) When two electrons are in opposite spin motion about the same axis, their net spin magnetic moment is zero.

Electron has another motion which is spin motion. Spin motion is rotation around itself as shown in figure 3.2.3 (a). If the electron is in spin motion, it also acquires a magnetic moment called spin magnetic moment which has the value,

$$\mu = \frac{-e}{2m} L$$

where L is the angular momentum of the electron and is its direction as shown in Fig. 3.2.3 (a). The spin motion of electrons is a major contributor to the magnetic properties of matter. So, it cannot be said that the electron in the atom may not have spin motion. As such there is room to say that the electron may not have orbital motion. Now if the net spin magnetic moment of the paired electrons remains zero, they must be in opposite spin motion about the same axis as shown in Figure 3.2.3 (b), which is physically possible. Because of this motion, the magnetic moment of both electrons is opposite and equal in value, so their effective magnetic movement will be zero, which is observed in the case of paired electrons in an atom. But it is necessary to see what this motion changes in terms of repulsion between them and this issue is not discussed yet. Obviously, there must be a difference in repulsive force between the electrons when they are stationary and when they are in spin motion about the same axis in opposite directions. Of course, when the electrons are stationary, the repulsive force should be maximum, when they come into opposite spin, the radial component of their electric field should decrease and the circular component should increase, thus reducing the repulsive force between them. The higher their spin velocity, the lower the repulsive force. This force will be a repulsive electric force but since the electron is in spin motion it will manifest itself as a tiny magnet as shown in Fig. 3.2.4 (a). Hence two electrons shown in Fig. 3.2.4 (b) in spin motion about the same axis will manifest as tiny magnets as shown in the figure in which their like magnetic poles are close to each other thus creating a magnetic repulsion between them. The higher the spin velocity of the electrons, the greater the magnetic repulsion they have.

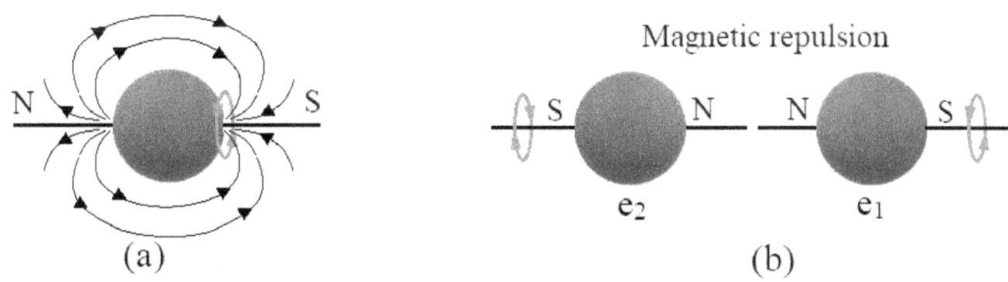

3.2.4 (a) Spin motion will cause the electron to manifest as a tiny magnet, (b) Spin motion in opposite direction will create magnetic repulsion between the two electrons.

It is important to see what happens if a positive core with +2e charge is placed at the center of such paired electrons as shown in Figure 3.2.5. Obviously, the positive core and each electron will have an electric attractive force. The higher the spin velocity of the electron, the lower the electric attractive force. Also, there will be electric repulsive force between both the electrons but it will decrease as the velocity increases. This effective electric repulsive force will be less than the effective attractive electric force. But at the same time, since the electrons are in spin motion and manifest as tiny magnets, they will have a magnetic repulsive force and it will increase as the spin velocity increases. Therefore, those electrons can be stabilized by balancing the effective repulsive force and the effective attractive force between them by maintaining a certain distance from the positive core and possessing a certain spin velocity. For this, the electrons will not need any orbital motion to settle at a certain distance from the positive core. In this the net effective electric charge becomes zero and also the net electric field and net magnetic field of this system which is necessary for the structure of the atom. This suggests that the spin motion is responsible for the structure of the atom. Of course, since this approach was not considered earlier, such an atomic model which can be called 'spin atomic model' could not come into being. The time at which different atomic models were proposed was influenced by how much was known about the spin motion of the electron, and the mechanics that were influential at that time also influenced the development of atomic models.

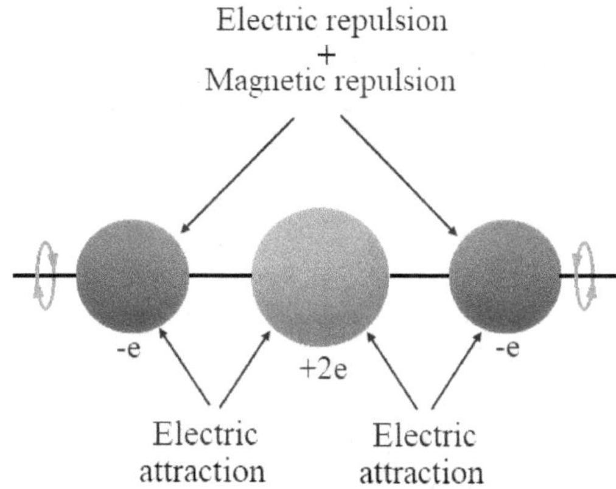

Figure 3.2.5. A positive core carrying +2e charge is placed between two electrons which are in opposite spin motion.

JJ Thomson proposed the plum pudding model of the object in 1887 after discovering cathode rays, i.e. electrons. Further, Rutherford showed in 1911 that there is a heavy positive charge at the center of the atom and around it are scattered light negatively charged particles called electrons. He proposed the planetary model of the atom and then Bohr proposed the quantum mechanical model of the hydrogen atom in 1913. It is because of this that the Planetary Model of Atom was also called the Rutherford-Bohr Model of Atom which is still popular today. Until then, no information was available about the spin of the electron. In fact, the spin motion of the electron is the major contributor to the magnetic property of matter, and although the magnetic property has been known since ancient times, we were unaware of the spin of the electron. In the 1920, Stern and Gerlach, followed by Gautsmith and Uhlenbeck and others, introduced information about electron spin, but did not consider whether electron spin contributed to the structure of the atom. Because at that time quantum mechanics was developing and due to de Broglie's hypothesis, an attempt was made to solve the problem of the atom by treating the electron as a wave, so physics began to move in a different direction. Although the above discussion shows that the spin motion of the electron is responsible for the binding of the atom. However, it is necessary to get answers to the questions of the minimum distance from the nucleus where two electrons form a pair, what their spin velocity should be, and how the forces between them are balanced. We know that electrons in an atom are fixed at different but fixed distances. So is that distance quantized? And why is that so? Is the spin responsible for it? Is the spin velocity of an electron constant? Many such questions arise. But it is a fact that electrons in an atom never move in orbit and it is only because of their spin motion that they can remain fixed at a certain distance from the nucleus. Another anomaly is that if spin is an intrinsic property of an electron, then its spin velocity would be fixed, and its magnetic moment would also be fixed. At that time there will be no change in its spin velocity, so how can their attraction and repulsion balance in the atom? To find out whether spin is an intrinsic property of electron or not, the experiment e/m by Thomson method can be considered.

3.3 Is spin motion an intrinsic property of an electron?

In 1921, Stern and Gerlach performed the experiment shown in Figure 3.3.1 (a) in which silver atoms split into two as they passed through a non-homogeneous magnetic field. This means that the atoms were expressing themselves as magnets in which some atoms had their unpaired electrons spin up while others had their spin down, thus splitting them apart. From this it was concluded that electrons have two types of spin motion, up and down. And further in 1925 Gautsmith and Uhlenbeck proposed that spin is an intrinsic property of electron. As shown in Figure 3.3.1 (b), one electron is up while the other electron is down. This means that the spin velocity and spin magnetic momentum will be constant and cannot change in any way, thus causing problems in the spin atomic model. To reduce the repulsion between two electrons, they are triggered in opposite spins and at the same time their magnetic repulsion increases and the attraction and repulsion balance between the paired electrons in the atom. If there is no change in spin velocity, this attraction and repulsion is difficult to balance over a certain distance and this process is difficult to happen. For that it is necessary to check whether the spin motion of an independent electron is an intrinsic property. If we look at it this way, the TVs that came into existence using CRTs could not have worked if spin had been an intrinsic property of electrons. This conclusion can also be drawn from the experiment e/m by Thomson method.

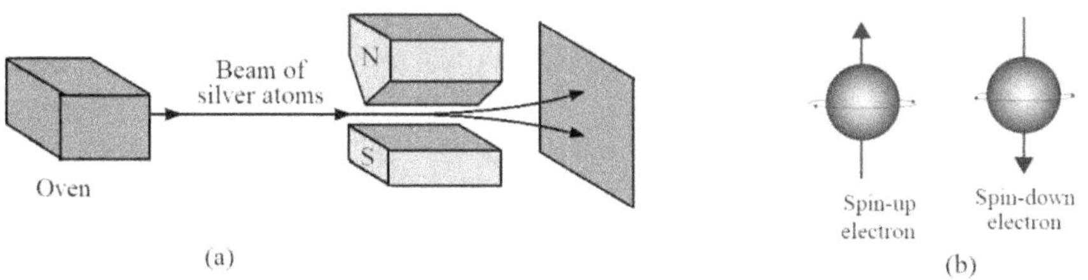

3.3.1. (a) Splitting of silver beam into two parts when passed through non-homogeneous magnetic field, (b) Electron spin up and down.

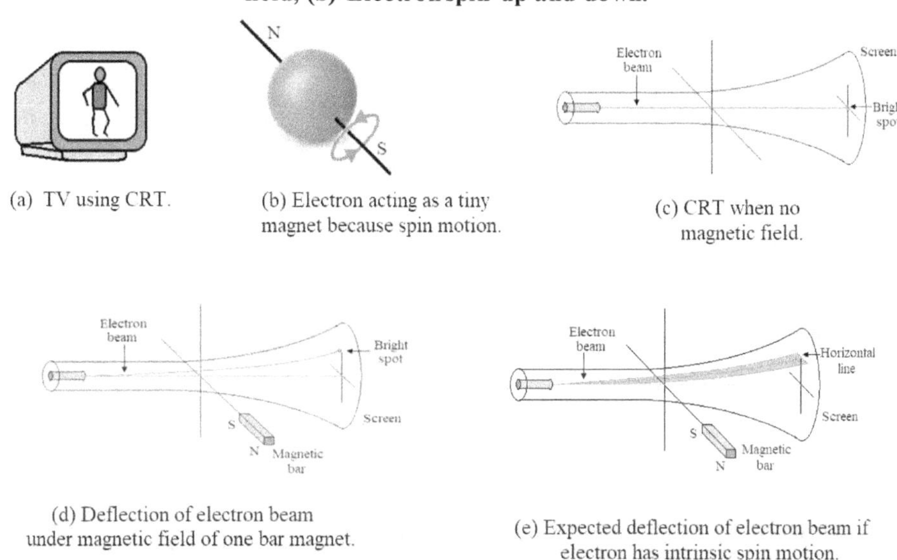

(a) TV using CRT.

(b) Electron acting as a tiny magnet because spin motion.

(c) CRT when no magnetic field.

(d) Deflection of electron beam under magnetic field of one bar magnet.

(e) Expected deflection of electron beam if electron has intrinsic spin motion.

Figure 3.3.2. (a) Television made using CRT, (b) Electron as a tiny magnet, (c) Deflection of an electron beam in a CRT when not in a magnetic field, (d) Deflection of an electron beam by placing a bar magnet, (e) If electron acts as a tiny magnet, the expected deflection of the electron beam.

An experiment proposed by Thomson to determine the charge to mass ratio (e/m) of an electron is carried out at undergraduate level in physics. We first observed that the beam of silver atoms was split into two parts in a non-uniform magnetic field. The main reason for this is the unpaired electrons in the silver atom whose spin motion manifests itself as a tiny magnet and that is why the silver atom gets its magnetic moment. But this does not mean that electron has intrinsic spin motion. Because if it is an intrinsic property of an electron, then the electron must remain in spin motion whether it is in the atom or not. So it is necessary to check the spin motion of the electron when it is outside the atom. An experiment by e/m by Thomson can be useful for that. If the electron is in spin motion as shown in Figure 3.3.2 (b), it will manifest itself as a magnet. So, in a non-uniform magnetic field, they will deflect, just as the silver atoms were deflected. A non-uniform magnetic field created by a single bar magnet can be used for this. Figure 3.3.2 (c) shows a CRT in which an electron beam has produced a bright spark in the center of the screen. Now a single magnetic bar is placed perpendicular to the electron beam as shown in Fig. 3.3.2 (d). Therefore, the electron beam is deflected upwards so that a spot A is formed of the screen. The non-uniform magnetic field created by using a single bar magnet linked to the electron beam of the CRT will apply a force on each electron in the electron beam according to the following law.

$$F = - e\,(v \times B)$$

where B is the magnetic induction in place of the electron and v is the velocity of the electron. In this equation, the electron is assumed to be only a charged particle. If the electron were assumed to have intrinsic spin motion, it would have expressed itself as a tiny magnet as shown in Fig. 3.3.2 (b). Since the magnetic field created by magnetic bars is non-uniform, electrons whose N pole is towards the bar magnet will be attracted towards the bar magnet and those whose S pole is towards the bar magnet will be pushed away from the bar magnet. So, a horizontal line would have formed at the top of the screen as shown in Figure 3.3.2 (d) but no such line appears there. This means that after the electrons leave the atoms, they give up their spin motion. That is, the question arises as to whether an electron acquires spin motion through the process of reducing repulsion with another electron in an atom. Of course, more research is needed on this.

3.4. Formation of Atoms

We have noticed that if two electrons form a pair and undergo spin motion in opposite directions, they can stabilize at a certain distance from the positive core. Maybe they have to adjust the spin velocity for that. This is why it is necessary to verify how different atoms can be built. Of course, the formation of atoms and molecules using spin motion is discussed further as a possibility. This model is yet to be refined.

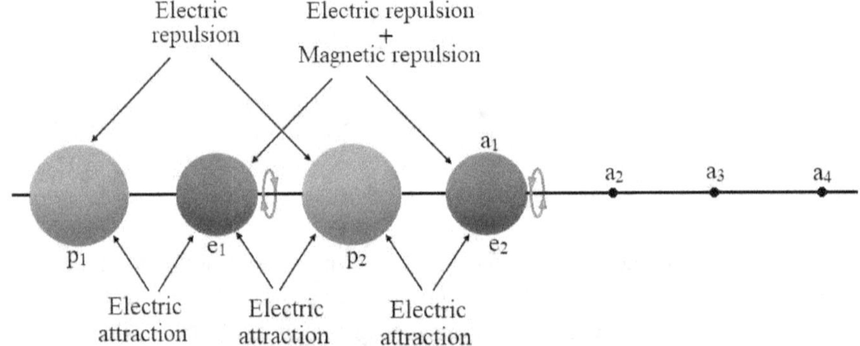

Figure 3.4.1. Formation of hydrogen molecule.

The first small atom is hydrogen, which has one electron and one proton, so the first problem lies in this atom. Since it does not have two electrons, there is no question of forming a pair, so in such a situation it is difficult to imagine how the structure of the hydrogen atom should be. Thus, Bohr has systematically analyzed the radiation spectrum of the hydrogen atom by assuming that the electrons move around the protons in certain orbits. But one thing we have to remember is that when an electron jumps from an outer orbit to an inner orbit as Bohr says, the difference in its energy will come out in the form of an electromagnetic wave, that wave will definitely have a certain number of crests and troughs. As many crests or troughs as there are, that electron in that hydrogen atom must oscillate at one place with the frequency of that wave. If the electron in the hydrogen atom is always moving in orbit, it will not be able to stay fixed in one place and oscillate, so the fundamental question arises as to how the electromagnetic wave is formed. So even in a hydrogen atom the electron should not move in the orbit and if it is true, it should form a pair to remain stable. The only alternative to this is to expect two hydrogen atoms to join together to form a pair of electrons between them and stabilize as shown in Figure 3.4.1. In this, there will be electric attraction between electron-proton and at the same time electric repulsion between proton-proton and electron-electron and magnetic repulsion between electron-electron. Both electrons will have the same spin velocity but in opposite direction and as it increases the electric force between them will decrease but the magnetic force will increase. By adjusting the distance between them and the spin velocity of the electrons, this system will be balanced and the hydrogen molecule H2 will be formed from it. The overall electric field and overall magnetic field of this system will be zero and the system will be stationary. Of course, what will be the distance between them and what will be their spin velocity, this conclusion could not be reached yet. If we consider the spectrum of hydrogen atom, a2 or a3 or a4 at certain distance where the effective attractive electric force and effective repulsive force will be balanced and the system will be transiently stable. At each such place when it takes a jump from another place, it will oscillate thus creating an electromagnetic wave. The frequency at which it oscillates must depend on the speed of the electron when it reaches that point. The farther the electron reaches that point, the higher its speed should be, i.e., the kinetic energy should be higher. When it reaches that point, the kinetic energy will be converted into potential energy and back to each other. So the electron will continue to oscillate and an electromagnetic wave will continue to be generated and the total energy of that electron will decrease and eventually it will come to zero. For this, mathematical equations of energy should be presented in a scientific manner and an attempt should be made to determine the frequency of the oscillations. The ultimate truth is that electromagnetic waves cannot be generated unless electrons oscillate at a point in space. The question is how can an electron in a hydrogen molecule remain fixed at different points (a1, a2, a3, a4,.......) as shown in Figure 3.4.1. Perhaps its spin velocity is quantized and hence it is transiently stable with different spin velocities at different points. At that time overall electric field and overall magnetic field will be zero. We need to find out if this spin velocity or the spin angular momentum is quantized.

The largest atom after hydrogen is helium, which typically has two protons and two neutrons in a nucleus surrounded by two electrons. Atoms are usually formed in the core of stars. Neutrons are believed to play an important role in binding protons together. But the role of spin of proton in the matter of nucleus formation and also exactly how nucleus is formed in atoms is still not clear. Also, it is necessary to see if the spin of the proton is related to the spin of the electron. We are currently keeping this topic aside from the discussion. Helium atoms are expected to form from hydrogen atoms. When one more proton and two neutrons accumulate in the nucleus of a hydrogen atom from the plasma, it will need one more electron to become electrically neutral and that electron will be taken from the plasma. But this electron will not fall directly into the nucleus. It will trigger into opposite spin motion to reduce electron-electron repulsion. This will reduce the electrical repulsion between the two but since the attraction between proton and electron is strong, the nucleus will try to pull both the electrons closer. At the same time, the spin velocity of both the electrons will increase and the magnetic repulsion between them will increase and the effective attractive force between electrons and nucleus and the effective repulsive force between electron-electron will be balanced at a certain distance from the nucleus of electrons and this system will be stable. The overall electric field and overall magnetic field of this system will be zero and the system will be completely stationary. A helium atom would thus form and have a structure as shown in Figure 3.4.2. These first two electrons will have the same energy so they can be considered electrons in energy level one. Also, the surface of the volume formed by them can be called the surface area of the first level in which the entry of further electrons is restricted.

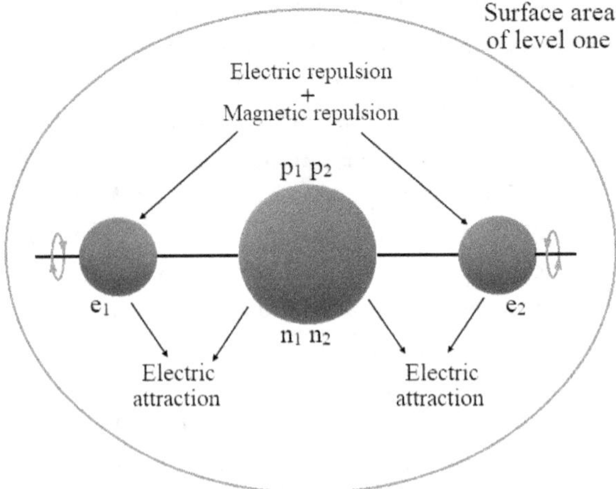

Figure 3.4.2. Structure of Helium Atom.

When another proton is deposited from the plasma into the nucleus, another electron will be drawn from the plasma to the system. But it will not be able to enter the volume of the first layer because the effective electric field of both the electrons in this layer will not allow this electron to enter. Both of these electrons are in spin motion so the electric field of each will also have a tangential or circular component. To reduce the repulsion between the two electrons in the first level and the third electron, the third electron will also go into spin motion with a spin velocity opposite to the effective spin velocity of both electrons in the first level. Its structure will be as shown in figure 2.4.3 called lithium atom. Since the overall magnetic field of this atom will not be zero, it will be unstable.

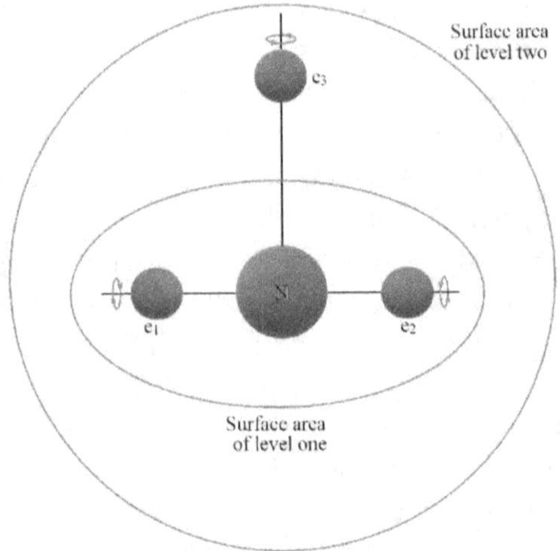

Figure 3.4.3. Structure of the Lithium Atom.

44

When another proton accumulates in the nucleus along with a neutron in this system, another electron will be taken from the plasma and balance the spin of the third electron. In this system the net electric field and net magnetic field will be zero and the system will be stationary and the system will be a beryllium atom as shown in figure 3.4.4. These two electrons will form a second level volume as shown in the figure.

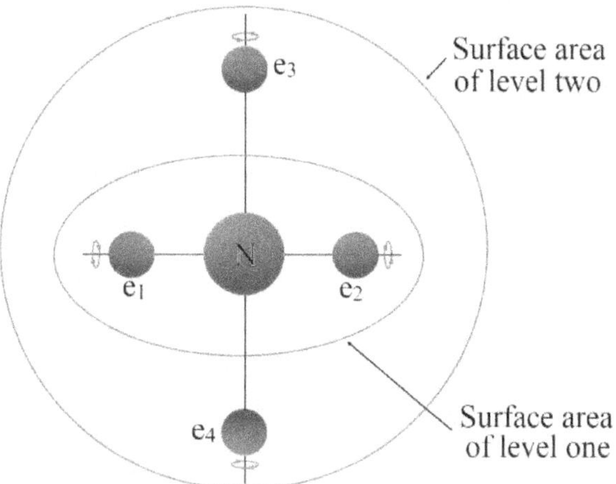

Figure 3.4.4. Structure of the beryllium atom.

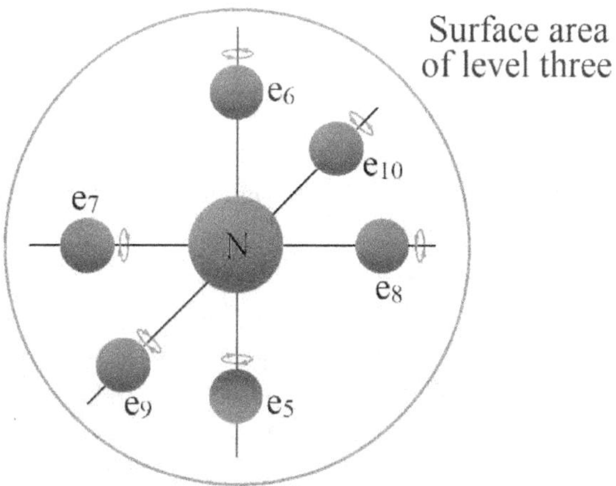

Figure 3.4.5. Structure of the Neon Atom.

Further when protons accumulate in the nucleus along with neutrons, electrons from the plasma will also be taken into the system to form a third layer of electrons. It appears that this level will be completed by an arrangement of six electrons in the third level, forming a neon atom as shown in Figure 3.4.5. The spin axes of these three pairs of electrons will be perpendicular to each other. Here the first and second levels of electrons are not shown in the diagram but are expected to be in the interior. The energy of electrons in different levels will be different but how to calculate is still a subject of research. Large atoms are thus formed by the accumulation of protons and neutrons in the nucleus and electrons in different outer layers. Of course,

45

how the electrons fit into the different levels by adjusting their spin motions is complicated. But it is true that electrons can form atoms through spin motion rather than orbital motion. There is a need to think deeply about this.

3.5. Formation of Molecules

In general, the total electric and total magnetic fields of each atom must be zero. Two or more atoms whose total electric and total magnetic fields cannot be zero can achieve this by coming together again. Due to this, a new joint system is formed and stabilized, it is called a molecule. Molecules are two or more unstable atoms that bind together to zero the total electric and magnetic fields and become stable.

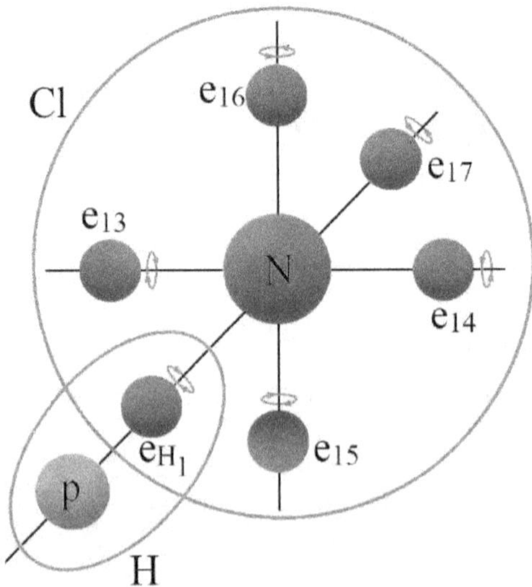

Figure 3.5.1. Structure of Hydrochloric Acid (HCL) Molecule.

We have seen that hydrogen is an unstable atom and how two hydrogen atoms combine to form H2 molecules. Similarly, we are going to discuss some specific molecules as examples. We will see how the HCL (hydrochloric acid) molecule, as shown in Figure 3.5.1, can be formed by combining a hydrogen atom and a chlorine atom. A chlorine atom has seventeen electrons. That means there are five electrons in the fifth level or shell. This means there are two pairs of electrons in this shell and one more electron is needed to form the third electron pair to complete the fifth shell of electrons. Also, a hydrogen atom has only one electron, so the two atoms can combine to meet each other's needs. For this, one electron of the hydrogen atom will be fixed in the fifth electron shell of the chlorine atom so that the unpaired electron of chlorine (e17) and the unpaired electron of hydrogen (eH1) form a pair as shown in the figure. Its total electric field and magnetic field will be zero and it will be a stable molecule. Here the chlorine atom gains an electron to complete its fifth level, therefore, it is the acceptor and the hydrogen atom is the donor.

Let us take the next example of a water molecule, which contains one oxygen atom and two hydrogen atoms, as shown in Figure 3.5.2. The third electron level of an oxygen atom contains four electrons, and two more electrons are needed to complete that level. It can fulfill its requirement by taking two hydrogen atoms. When an oxygen atom pulls two hydrogen atoms close together, the naturally adjacent one will first pair with the oxygen electron (e7) and then pair with the remaining unpaired electron (e8) as the other hydrogen atom approaches. Thus, the oxygen atom will complete its third electron shell. The total electric field and magnetic field of this molecule will be zero and the molecule will be stationary as shown in Figure 3.5.2. An oxygen atom is an acceptor because it takes two electrons from two hydrogen atoms to complete the third level. Hence the hydrogen atom will be the donor.

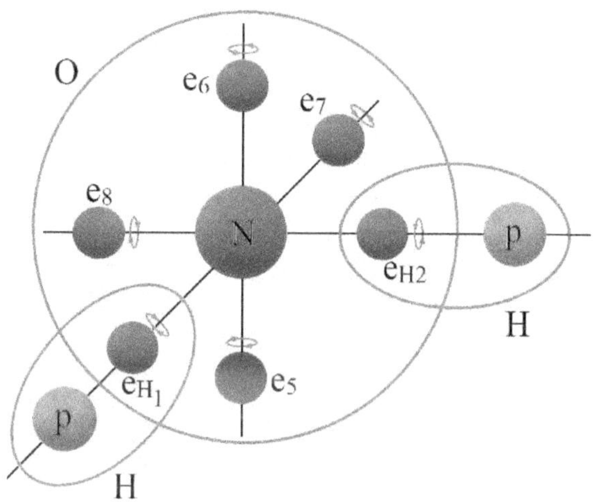

Figure 3.5.2. Structure of water molecule.

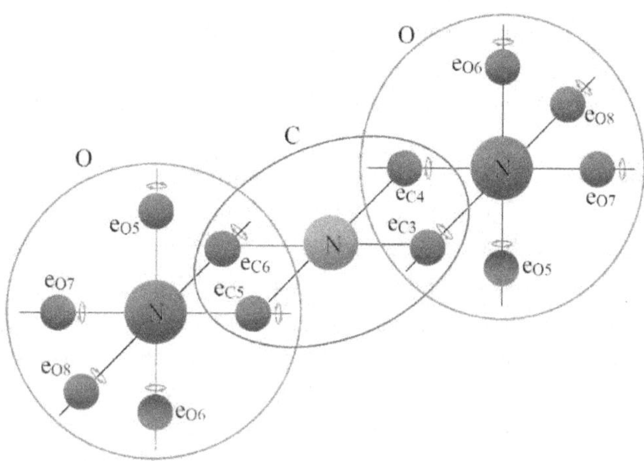

Figure 3.5.3. Structure of carbon dioxide molecule.

If a more complex molecule is to be discussed, one can look at the carbon dioxide molecule as illustrated in Figure 3.5.3. A carbon atom has two electrons in the second electron shell (eC3, eC4) and two electrons in the third shell (eC5, eC6). Oxygen has four electrons in its third electron shell. Carbon's two electrons (eC3,

47

eC4) align with the oxygen atom's two unpaired electrons (eO7, eO8) to form two pairs. Also, another oxygen atom can form two pairs with the other two electrons (eC5, eC6) of the carbon atom. Thus, the two oxygen atoms accept four electrons from the carbon atom to complete their third electron shell. So here oxygen atoms are acceptors and carbon atoms are donors. The total electric field and magnetic field of this molecule is zero, so it is a stable molecule. There are many possibilities for the formation of many such molecules. And this is possible only because the electrons do not move in orbits in the atom. They stabilize by forming pairs in atoms. This new and realistic approach required a re-understanding of the chemical formula of each molecule.

It is true that electrons in an atom do not move in orbits. Therefore, it is wrong to show the electrons moving in orbitals when showing the atomic model, which reinforces the misconception that electrons move in atomic orbitals. The spin motion of an electron is a physical motion responsible for the formation of atoms and molecules. Hence this model of atom can be considered as 'spin atomic model'. It should be noted that if the electrons in the atom had remained in orbital motion, their effective magnetic momentum could never have been zero and a solid material of fixed size could not have been formed. However, in this model, it is not yet understood how the electrons are distributed in the different quantized levels. It also remains to calculate the energies of the electrons in the different levels which will verify the emission and absorption spectra of the atoms.

3.6 Summary

1. The idea that electrons move in orbits around the nucleus in an atom is derived from planetary motion. But if the electrons are moving in the orbit, they will continue to radiate energy and eventually fall into the nucleus but this is not observed, hence the concept of stationary orbit was introduced. In which the electrons will not radiate energy as they move. In fact, it makes no sense to say that electrons do not radiate energy while moving in a stationary orbit. It is not on the electron's mind.

2. It is a fact and accepted by all that electrons are in pairs in atoms and their net magnetic moment is zero. They both have the same energy which means they must both be moving in the same orbit and that too in opposite directions which is never possible. Because they have repulsion but their orbital magnetic moment cannot be zero unless they move in the same orbit and in opposite directions. This means they do not move in orbit.

3. Electrons in an atom have spin motion and due to this motion of unpaired electrons that the substance acquires its magnetic property. It means that spin motion of electronic is a physical motion.

4. Paired electrons in an atom have zero effective spin magnetic moment, which means they must be in spin motion opposite to each other on the same axis, but this will create electron-electron magnetic repulsion. And when the effective attractive force between the nucleus and electrons and the effective repulsive force between electron-electron are balanced at a certain distance of the electrons from the nucleus and at a certain spin velocity of the electron, the electrons can remain stable at a certain distance from the nucleus in the atom. This does not require them to have orbital motion, nor is it possible.

5. If all the electrons in an atom were moving in orbit, it would be impossible to form a solid material with constant shape from such atoms.

6. Many atoms emit light and such a light wave has many numbers of crests and troughs. Suppose a light wave has n number of crests and troughs, then the electron in the atom, due to the transaction of which this wave is formed, must vibrate n times at one place, otherwise such a wave cannot be formed. For that, the

electron should not be moving in orbit or according to quantum mechanics, even if the electron is expressed as a wave, an electromagnetic wave cannot be generated from it.

7. In fact, no one yet knows, and no one seems to be talking about, how the energy difference of an electron is converted into an electromagnetic wave.

8. The electron's spin is not its interesting property, since TVs would not have been produced using CRTs. Also, the e/m experiment shows that the electrons in the beam of CRT have no spin motion. This means that electrons possess spin motion while inside the atoms and their spin motion disappears after they come out of the atoms.

9. The energies of the electrons still remain to be deduced by assuming spin motion in the atom. Also, the equation of the frequency at which it vibrates after jumping from one place to another is yet to be worked out which is very important. Also, how spin motion stabilizes electrons at different distances in an atom and how different electron layers are formed is yet to be understood. Electrons having the same energy should be in the same layer.

10. If an atom has unpaired electrons, it forms a pair with an unpaired electron from another element to form a molecule and become stable.

11. It is true that electrons do not move in orbits. But in the logos of various Atomic Energy Agencies and in many media, the electrons in the atom are shown moving in orbits, so this belief is rooted in the students which is wrong. What we do not know for sure is not known and therefore should be put before the world so that the youth can become aware of it and express their thoughts and research about it.

Chapter - 04

What is a magnetic field?

4.1 Historical background of magnetic field

Nowadays no one seems to question what a magnetic field is. I don't know how basic research has been so messed up. In Indian Hinduism, the practice of meditation has been of extraordinary importance since ancient times. Meditation is the contemplation of the universe in solitude. Detached from the practical, the attempt to find the answer to the question 'Who am I?' is meditation practice. Where did this universe come from? In fact, there is no reason for the creation of this universe. So, did anyone knowingly create it? Is there a manufacturer of this? Is there a driver for this? What is my purpose? Attempting to track these is a meditation practice. Of course, it also contemplates how and why the circulation of universal things takes place. This is the spiritual method of Hinduism to track the universe. Another scientific method of tracking the universe is science, which involves drawing conclusions based on logic, testing them through experiments, and confirming such logical statements. Spirituality comes to the conclusion that the universe is filled with a certain consciousness, which is the cause of the creation of the universe, also known as Paramatma. That is, every element in the universe is a part of God, in fact it is made up of God. Science also believes that the universe must be filled with a specific medium and that is the ether which is responsible for the generation and propagation of electromagnetic waves and also for the existence of electric and magnetic fields. Not only that, but he must be responsible for the creation of the substance. But unfortunately, we have not been able to prove its existence yet. As a result of this, the time has come for us to make different rules for every event that happens in this universe. In fact, there must be some single thing in the universe that must be the cause of every event that happens, but we couldn't determine that, and each event seemed independent. It cannot be related to another phenomenon, so it is time to make a new rule for each phenomenon. This has led us to believe, in fact to misunderstand, that the universe behaves differently at different levels, making understanding the universe even more complicated. Things seen in normal daily life also began to appear in different ways. This is also due to the effort made to tie the phenomenon into a mathematical equation rather than finding an answer as to why the phenomenon occurs. In fact, if we want to understand the universe, it is necessary to find the answer to 'why this?', not to frame it in equations. One such thing that we experience in our daily life is the magnetic field. Is the magnetic field real or an illusion, or is there something else? This also never seems to have been discussed. Magnetic monopoles do not exist in the universe, then what is the magnetic field? Human beings should not be allowed to sit cheaply on this issue. Just as electric monopoles i.e., electrons and protons exist, magnetic monopoles of S and N must exist in the universe, otherwise the reality of a magnetic field can be questioned. It is amazing that no one doubts that we have accepted the existence of magnetic fields even though magnetic monopoles do not exist in the universe. If the existence of a magnetic field is not real, and yet we assume that its existence is real, there must be some anomaly in the nature of its effect. But we should be able to look at such inconsistencies with an eye. There are so many anomalies in the magnetic field that we have never looked at that way. Next, we are taking some of them for discussion.

50

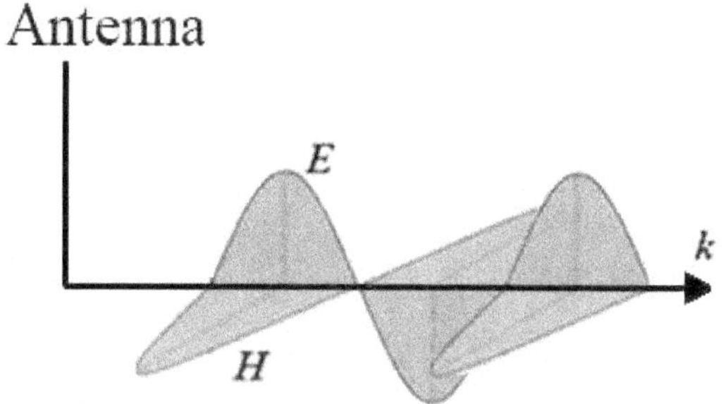

Figure 4.1.1. An electromagnetic wave produced by an antenna.

As shown in Figure 4.1.1, let us create an electromagnetic wave through an antenna that propagates with speed c and contains electric and magnetic fields. Now, even if the alternating electric current in the antenna, which is generating the electromagnetic wave, is turned off, the generated wave will continue to propagate which contains electric and magnetic fields. Now this wave is independent and existence of electric field and magnetic field in it is real and not imaginary. When this wave interacts with a charged particle, its electric and magnetic fields will exert a force on the particle. Now passing an electric current through a straight conductor produces a circular magnetic field as shown in figure 4.1.2.

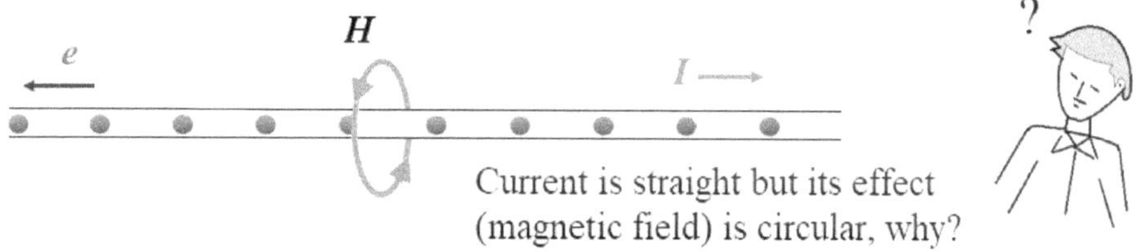

Figure 4.1.2. A magnetic field created from an electric current.

Now this magnetic field is also real which means it is not imaginary. The reason it is not imaginary is that if a charged particle is moving along that conductor, the effect caused by that current is to apply a force according to the law of the magnetic field. This magnetic field is the effect of the current in that conductor. Current is straight so why its effect is circular? This simple but very important question does not occur to anyone and no one can rationally support it. Does this mean something is wrong with the magnetic field we assume? Or do we still not understand it properly? One wonders why such a basic question does not arise in the minds of researchers.

Another example is Faraday's Law of Electromagnetic Induction. The emf produced in an electric circuit is equal to the rate of change with time of the total magnetic flux ($\emptyset$) linked to that circuit. That is,

$$emf = -\frac{d\emptyset}{dt}$$

Of course, the current generated in that circuit depends on the rate of change with time of the total magnetic flux linked to that circuit, and that current is the motion of the electrons in that circuit. In fact, the equation for the magnetic force is,

$$F = q(E + v \times B)$$

where, q is the electric charge on the particle and v is the velocity of the particle and B is the magnetic induction at the position of the particle.

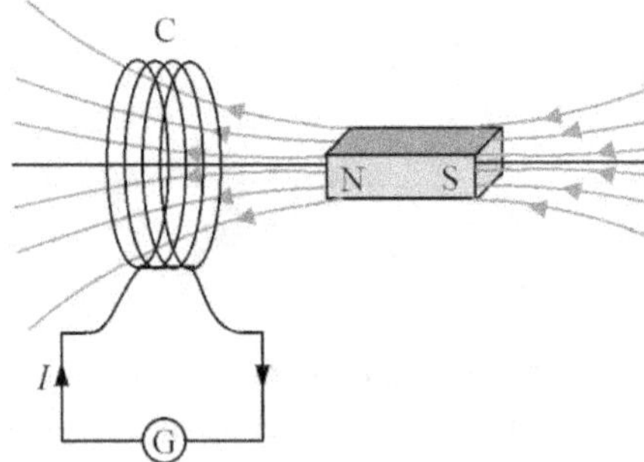

Figure 4.1.3. According to Faraday's Law of Electromagnetic Induction, current is created in a closed circuit.

This is the only equation of the magnetic force that describes how a magnetic field exerts a force on any electrically charged particle. Then why this equation cannot explain electromagnetic induction. We do not have a suitable answer as to why it is time to make a separate and new rule for that. This means that we still do not understand the magnetic field very well or properly.

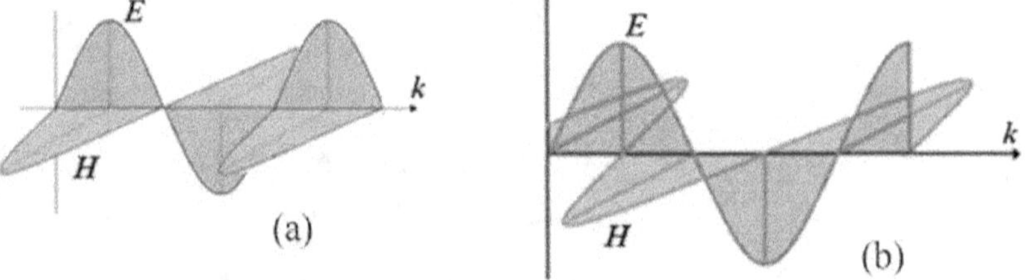

(a) According to the classical electrodynamics, electric and magnetic field waves are in phase.
(b) According to the explanation of Photoelectric effect, expected phase difference of 90 degree in electric and magnetic field waves.

Figure 4.1.4. Phase difference between electric and magnetic field wave expected (a) according to classical electrodynamics, (b) effect of asymmetric electric field.

A third example is what we have seen in explanations of the photoelectric effect, in which as the frequency of an electromagnetic wave increases, its magnetic force increases, and there is a 90-degree phase difference between the electric field and the magnetic field in that wave, and this cannot be explained by current electrodynamics. This clearly means that the magnetic field is not yet understood by us and its true nature is going to be difficult to discover as it has not yet been observed. It is also very difficult to decide in which direction to move. Now electric field is a kind of pressure created in space, pressure created around positively charged particles can be considered as inward pressure and around negatively charged particles can be called outward pressure. That is, one is a positive electric field and the other is a negative electric field. If electric pressure is to be produced, an elastic medium is required, which is ether. Unfortunately, we have not yet discovered the existence of ether. Naturally it will support only one field which is the electric field. This means that no magnetic field or gravitational field will exist in this ether. That is, some type or property of the electric field will show the effect of the magnetic field. Now it's not easy for anyone to figure out. One would think that it must be an electric field. Perhaps for this he would first examine the attraction and repulsion between current carrying wires and for that first try to find out what kind of field there is around the wire. For that, he will study the deflections of the electron beam in the CRT at different angles near the current carrying conductor for a few days but nothing will come up. He might be disappointed. After a few days he will feel that there is something he does not understand. Because his head will be influenced by prevailing electrodynamics. So even if he thinks it must be an electric field, not a magnetic field, he won't find a way out any time soon. And one night suddenly the tube in his head will light up and he will begin to think that there might be some kind of possibility. That will make him impatient. The next day he will go to the lab and do some experiments and come to a conclusion. Actually, he didn't need to do the experiment, he knew what was going to happen but he wanted to do the experiment from a different perspective.

4.2 Understanding the true nature of magnetic fields

The simplest experiment to make a preliminary estimate of what the magnetic field is is to pass a direct current through a straight conductor and examine the field produced around it with a cathode ray tube (CRT). This was the only option available. Actually, the correct option was to check its movement by placing a single electrically charged particle i.e. an electron near the current, but as that was not possible, he had to use another option, a CRT. By passing a direct current through a straight wire, the deflection of the electron beam in the CRT in different directions as shown in figure 4.2.1 was investigated but there was no trace of the electric field.

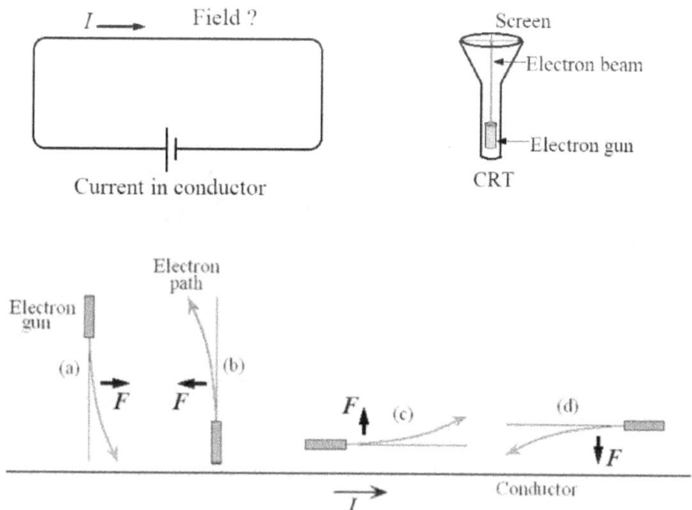

Deflections of electron beam near a straight long
conductor carrying current.

Figure 4.2.1. Deflections of electron beam in CRT near a current.

In figure 4.2.1, direct current is passing through the wire and deflections are shown when the CRT is placed in different directions near it. In (a) the CRT is placed perpendicular to the wire where the electrons are deflected to the right as they approach the wire which means the force on them is to the right. This suggests that there may be an electric field parallel to the wire. But when the CRT is rotated in the opposite direction, the deflection appears to the left as shown in (b). So, these are not signs of an electric field. Because even if the direction of the charged particle changes in the electric field, the direction of the force does not change. Similarly, in (c) and (d) the direction of force is reversed. This clearly shows that there is no electric field around this wire. In fact, it was a headache. But this question did not cross his mind. Then he realized that this force must come from field-field interaction, which means that our conventional equation of electric force $F=qE$ does not apply. Because the parallel electric field, which decreases away from the wire, applies an asymmetric force on the electron, the electron is deflected instead of going straight. But to understand this force, the electron had to be divided into two parts, and since the electron is a point, it was not digestible to anyone. Yet there was no dispute that the electron was deflected, meaning that the force was asymmetric, and then he was realized that this force must arise from field-field interaction, which we have already noted in the photoelectric effect. That is, this thing can be understood only if the electric field parallel to the current applies a force on the field of electrons. Since the electric force arises from field-field interaction, some conclusions can be reached. Of course, he knows that the magnetic field puzzle is not that easy to solve, otherwise it would have been solved long ago. By considering these things, we can try to solve the magnetic field puzzle as follows.

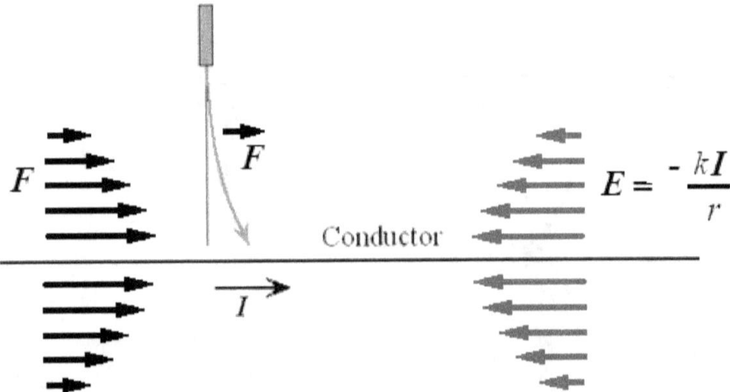

Figure 4.2.2. When a direct current passes through a long wire, the possibility of creating an electric field parallel to the wire.

First it is necessary to see if the current in a straight wire can produce a parallel electric field around the wire. It can form because the current causes only the electrons in the wire to move, so their electrical tangential component cannot balance because the positive charges are not in motion. The parallel electric field which is in opposite direction of current is effective residual component of electric field of the current. Its equation would be

$$E = - \frac{kI}{r}$$

where r is the perpendicular distance from the wire at the point of electric field and I is the current in the wire.

Now one thing is clear from this that the electric field of the conductor is not symmetric around the electron in the CRT so an asymmetric electric force will exist and hence the electron will always follow a curved path. To understand this let us divide the electric field of the electron into two parts, the lower electric field E_a and the upper electric field E_b. Now the force applied by the field of the wire on the field of the electron can be divided into two parts, the force F_a will be greater than the force F_b, so the electron will be pushed towards the force as shown in Fig. 4.2.3(a) as it approaches the wire. As shown in figure 4.2.3(b), even when the electron moves away from the wire, the force F_a is greater than the force F_b, so the electron must be pushed in the direction of F_a, but because the electron has momentum, the electron appears to move in the direction opposite to the effective force. Also, when an electron moves parallel to the wire, when it accelerates, it will naturally be pushed into the weak field i.e., away from the wire and when it decelerates, it will be pushed towards the strong field i.e., pulled towards the wire.

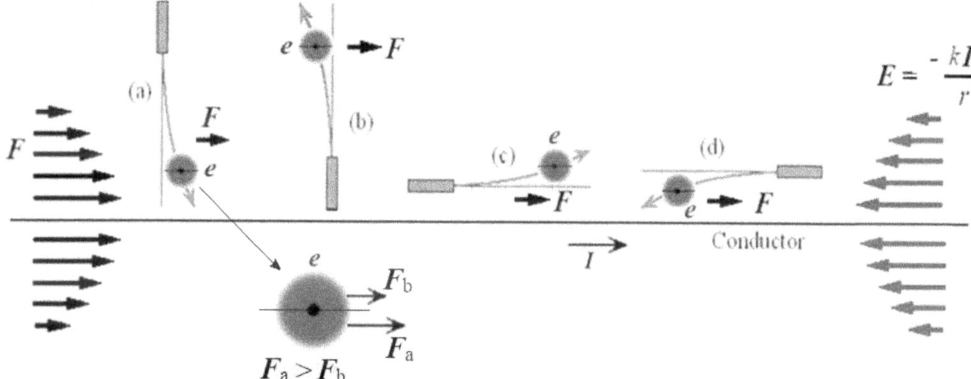

Figure 4.2.3. Explanation of deflection of electron beam in CRT when direct current is passed through wire through field-field interaction.

In fact, when force comes into existence acceleration occurs i.e., change in speed or change in direction, then it is very important to find out whether this force changes direction by changing speed or only change in direction due to asymmetric force. Our understanding is that the magnetic field only changes the direction of the particle, it does not change the speed of the particle. But here it seems that there should be a change in the direction of the electron as well as the speed. So, solving the puzzle of magnetic field is not so easy. Indeed, we should admire the wisdom of whoever solves this completely. For now, though, we assume that an asymmetric electric field has no effect on a stationary charged particle, but if the particle is in motion, it changes its direction as shown above. But in the photoelectric effect we are the first to see both happening. Because in the universe we find that if there is a force on an object in a field, that force tries to pull that object from a higher potential to a lower potential. At that time the speed of the object increases. Now if we pass a direct current through a circular coil as mentioned above, a circular electric field will be created as shown in the figure 4.2.4. But here there is no potential difference which means that there is no question of electron being pulled from higher potential to lower potential. So, is it just the electron changing direction? If so, then this new information is coming to light. But a potential difference is required to create a field. So, what is the parallel electric field around the wire? An electromagnetic wave has both a parallel electric field and a potential difference, as the ether is pulled from one point to another, creating a potential difference. That is why in the photoelectric effect the electron accelerates and its direction also changes. So, the question arises whether the field with potential difference is called electric field and the field without potential difference is called magnetic field. It gets even more complicated. But sometimes the puzzle of the magnetic field must be solved, and perhaps only through a collective effort. But one thing is certain that there is no magnetic field as we consider it. Because conventional electrodynamics cannot explain why the magnetic force increases when the frequency of an electromagnetic wave is increased and also why there is a 90-degree phase difference between the electric and magnetic fields in that wave.

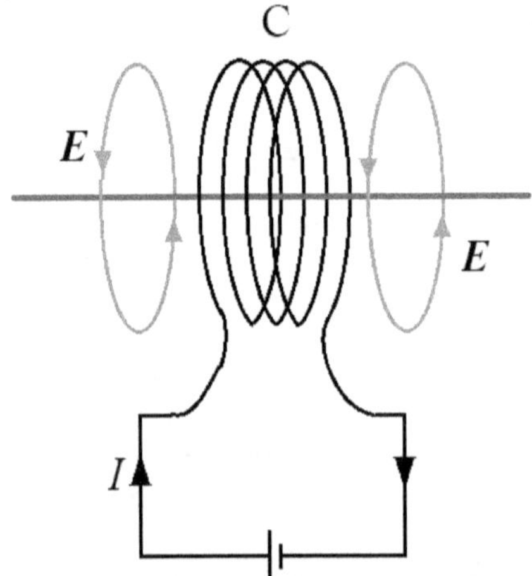

Figure 4.2.4. Electric field of a coil.

In fact, the concept of magnetism came from the properties of bar magnets. Since ancient times, magnetic needle was used as a guide. Two dissimilar magnetic poles attract each other and two homogeneous poles repel each other. The equation of force between them is the same as the force between two electrically charged particles, so it follows that just as there exist two types of electric charges, positive and negative, there must also exist two types of magnetic poles S and N. As the concept of electric field has come forward, the concept of magnetic field has also come into existence. The only difference is that electric charges exist but magnetic monopoles do not. The fact that the universe is perfect begs the question why magnetic monopoles do not exist, and if they do not, why magnetic fields exist. Or we do not yet know its true nature which we are trying to understand. The above experiment indicates that if an electric current flows through a straight and long conductor, a parallel electric field is created around it whose direction is opposite to the direction of the current. All the interactions of magnetism should be explained from this one clue. One thing to remember in this is that the electric field is symmetric and its curl is not zero and in such a field when an electron is in a position to accelerate it is pushed towards a weak field and when it is in a position to decelerate it is pulled towards a strong field.

4.3 Attraction and repulsion between two parallel conductors or wires when direct current is passed through them

As shown in Fig. 4.3.1 (a), a direct current I_a flows through a conductor A, creating a parallel electric field E_a around it which is opposite to its current and naturally decreases with distance from the conductor. A conductor B is placed in the field through which a current I_b is flowing. The electrons in it are in a position to be accelerated by the electrical field E_a so these electrons will be drawn towards the strong field of E_a i.e., towards conductor A. Also, conductor B will produce a parallel electric field E_b due to the current I_b in it which is in a position to decelerate the electrons flowing in conductor A. So, these electrons will be attracted to the strong field of E_b i.e., towards conductor B. Due to these mutual interactions, both the conductors will be attracted towards each other, when we pass direct current through two parallel conductors, we say that attraction is created between them. As shown in Figure 4.3.1 (b), if current flows through both conductors in opposite directions, they will repel each other. In this the electric field of conductor A is in a position to accelerate the electrons flowing in conductor B, pushing them towards the weaker field of Ea. Hence conductor B will be pushed away from conductor A. Similarly, conductor B will also react with conductor A thus creating mutual repulsion between them.

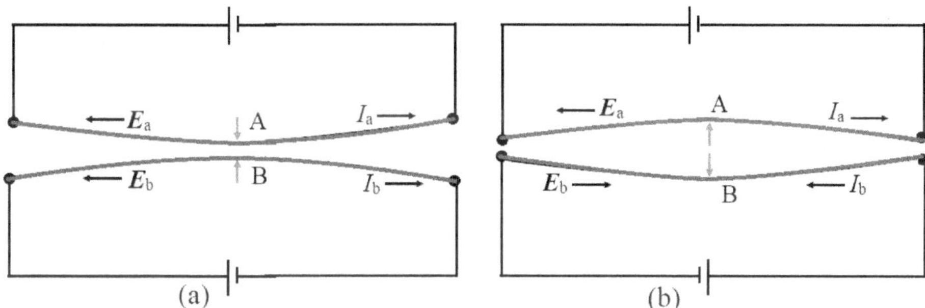

Figure 4.3.1. Attraction and resistance between two parallel conductors when direct current flows through them.

Actually, one thing that gets confusing here is that suppose there is an electric current I_a in the circuit A as shown in figure 4.3.2 and it creates a parallel electric field E_a around the conductor A. Another closed circuit is placed next to it with no power supply connected and its conductor B is parallel to conductor A. The electric field E_a is linked to the conductor B so it must apply a force on the electrons in the conductor B. So current should flow in circuit B in the opposite direction to the current in circuit A but it does not seem to be happening. But if the circuit B is made of superconductor, then definitely the expected current flows in it. So, it needs more clarity and we will discuss the matter in detail later in Electromagnetic Induction Produced in Secondary Circuit. In fact, it is still much more necessary to understand the behavior of charged particles in electric fields, symmetric and asymmetric.

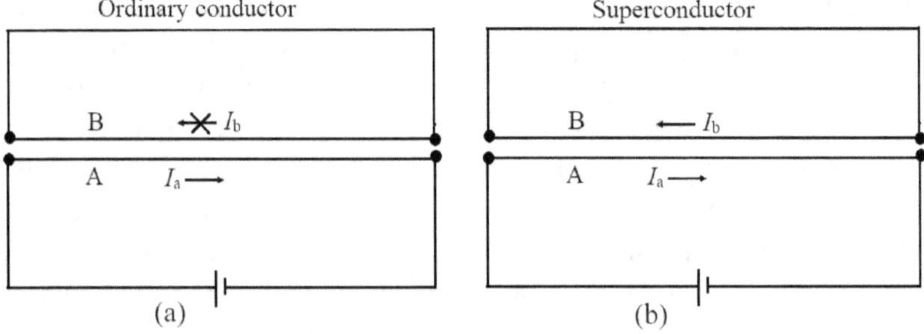

Figure 4.3.2. EMF produced in secondary circuit (a) not in ordinary conductor, but (b) in superconductor.

4.4 Possible electric field of a coil and bar magnet

As discussed earlier, if a direct current produces a parallel electric field, then obviously a circular coil should produce a circular electric field as shown in Figure 4.4.1. A circular electric field should be created which should decrease as it moves away from the coil. In this the area around the coil can be divided into three sections, Section A, Section B and Section C. In all three sections the direction of the circular electric field will be opposite to the current I and naturally in all three sections the electric field strength will decrease as we move away from the coil. In section A, the electric field strength should be higher near the conductors of the coil. And it should decrease as we move towards the axis of the coil. Also, it should decrease moving away from the coil in the direction of the axis. Also, in region B the field strength should decrease away from the coil. This means that in between region A and region B must have an electric field cone on surface of which the electric field is maximum. Similarly, a field cone will be formed between region B and region C. As one moves away from the coil in any direction, the strength of the electric field decreases. Now we have noticed that where the magnetic lines turn, the surface of electric field cone lies. Interestingly, the direction of the electric field in both the field cones is the same, so there is no reason to call them S pole and N pole as both the field cones are the same.

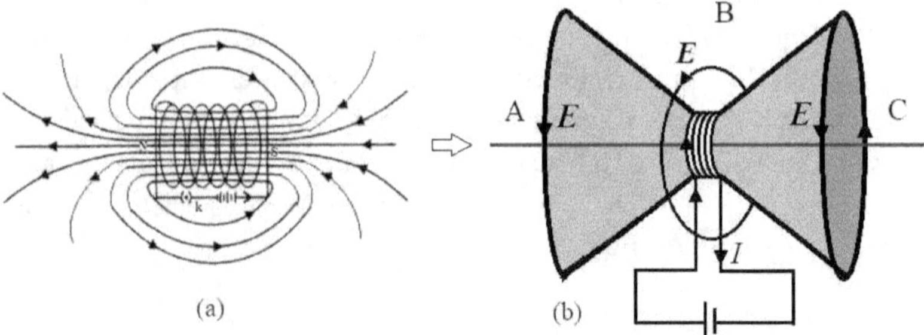

Figure 4.4.1. (a) Prevailing magnetic field of the coil, (b) Expected asymmetric or circular electric field.

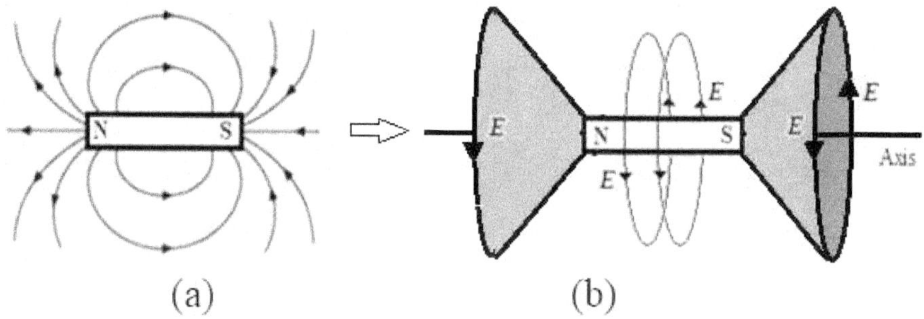

Figure 4.4.2. (a) Prevailing magnetic field, (b) Expected asymmetric or circular electric field of a bar magnet.

The magnetic field of a bar magnet and the magnetic field of a coil are almost the same, so instead of the magnetic field, their asymmetric electric fields will be as shown in the figure. Bar magnets contain unpaired electrons. Their spin motion aligns and an effective circular current is generated from them as shown in Figure 4.4.2 (b) which causes the formation of an asymmetric or circular electric field which is in the opposite direction of the current. Also, this field will have two cones as shown in figure. There is no reason to say that every bar magnet has S pole or N pole as both cones are the same. This suggests asymmetric electric fields or circular electric fields instead of magnetic fields of different planets. In the absence of solar wind, Earth's asymmetric electric field should be as shown in Figure 4.4.3.

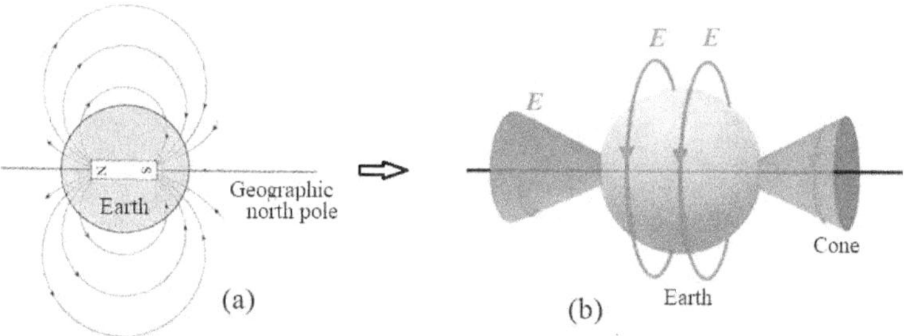

Figure 4.4.3. Earth's (a) prevailing magnetic field, (b) expected asymmetric or circular electric field.

Here too there will be two cones at whose surface the electric field is relatively maximum. The actual cause of the formation of these two electric cones is the shape of the aurora formed above both poles of the earth which is in the shape of a circular ring as shown in figure 4.4.4(a). This means that the activity of aurora formation is higher on the surface of the cone, because that is where the strength of the electric field is higher. Figure 4.4.4(b) shows the image of the aurora forming over the planet Saturn and that too in the form of a circular ring. If the prevailing magnetic field had produced the aurora, it would have formed in the form of a circular spot as shown in Figure 4.4.4 (c), because the density of the magnetic field lines is higher above the axes passing through the poles.

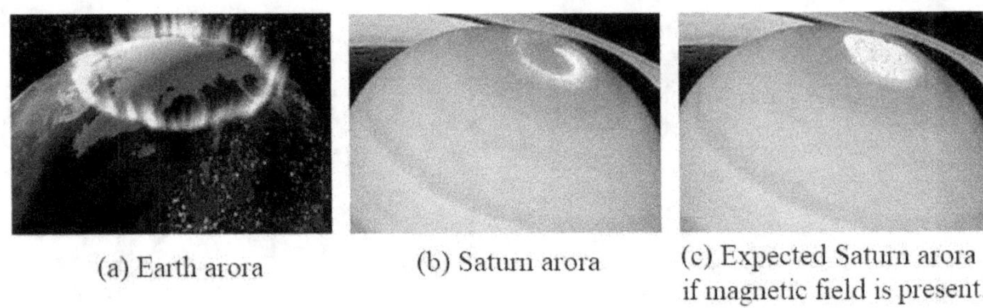

(a) Earth arora   (b) Saturn arora   (c) Expected Saturn arora
if magnetic field is present.

Figure 4.4.4. (a) Aurora seen at Earth's poles, (b) Aurora seen at Saturn's poles, (c) Expected aurora at Saturn's poles given the prevailing magnetic field.

4.5 Confirmation of the electric field cone of a bar magnet

Figure 4.5.1 (a) shows the circular electric field in region A and region B of a bar magnet. In both the regions the direction of the electric field is same but in region A the electric field strength decreases further away from the bar magnet as shown in figure 4.5.1 (b). Also, in region B the electric field strength increases from the axis to the surface of the angle and reaches a maximum at the surface as shown in Fig. 4.5.1 (c). As shown in figure 4.5.1 (a) the direction of circular electric field in both the regions is the same, the deflection of the charged particles will be seen to be opposite to each other which can be explained by figure 4.5.1 (b) and (c). In Figure 4.5.1 (b) a charged particle is moving towards the center of the field with speed v but it appears to be deflected upwards. By studying its motion at three points in this field one can conclude that the deflection that occurs will be due to the asymmetric nature of the electric field and will only change the direction of the particle. At position one, the charged particle will have a stronger force on the front electric field than on the back electric field, so it will move towards the electric field.

60

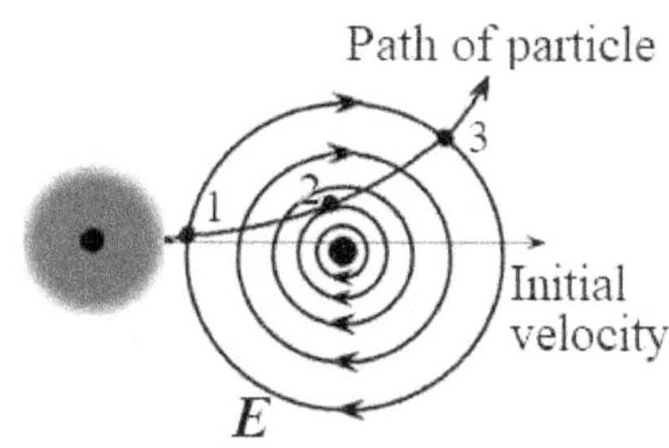

(b) Electric field decreases away from center as like in the region 'A' of the bar magnet.

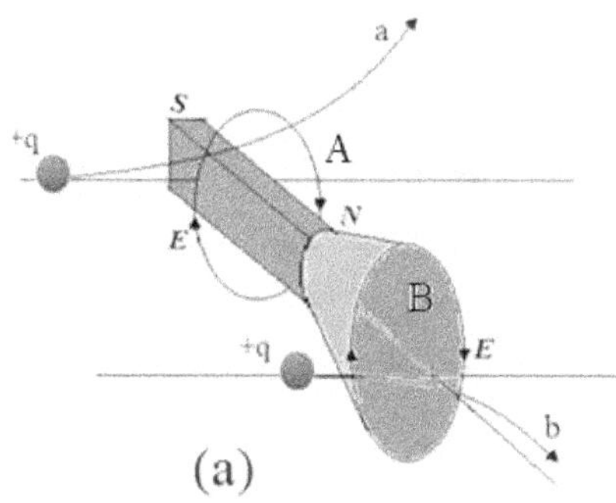

(a)

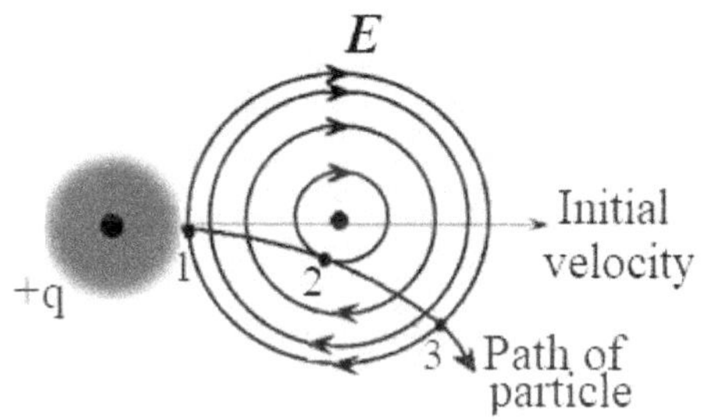

(c) Electric field increases away from center as like in the region of cone of the bar magnet.

Figure 4.5. Deflection of a charged particle near a bar magnet explained by a circular electric field.

At position two it is in a position to accelerate so that it moves towards the weaker region of the circular electric field, i.e., further upwards. At position three, the force on the lower electric field is greater than the upper electric field on the particle, so it will move further to the left. The electric field in region C is shown in Fig. 4.5.1 (c) in which the particle is moving with an initial velocity v but appears to be deflected downwards instead of going straight. Here also conclusions can be drawn by studying the motion of the particle at three places in the circular field. At position one, the force on the particle is stronger in the back electric field than in the front electric field, so the particle will move to the lower side. As the particle is decelerating in position two, it will turn towards the strong field i.e., downwards, and in position three, the force above the front electric field is stronger than the electric field behind the particle, so it will turn further down. Overall, the particle will follow a curved path. Although the direction of the electric field in Fig. 4.5.1 (b) and (c) is the same, the particle deflections in both are opposite to each other. This means that in a circular electric field, whether the electric field strength is increasing or decreasing from the center will determine which direction the particle will move. These deflections confirm the shape of the bar magnet's circular electric field. This does not require the bar magnet to assume S and N poles.

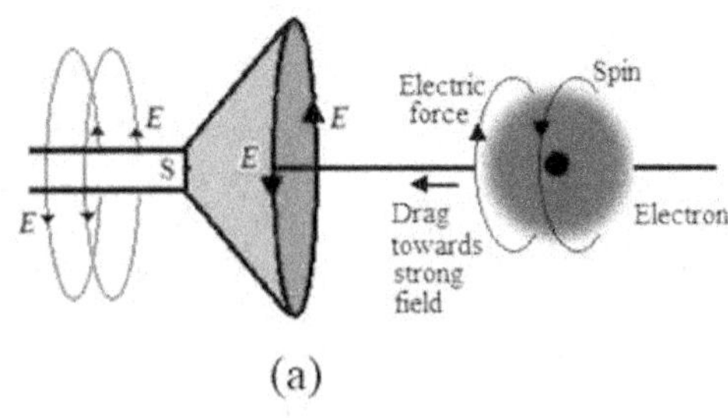

(a)

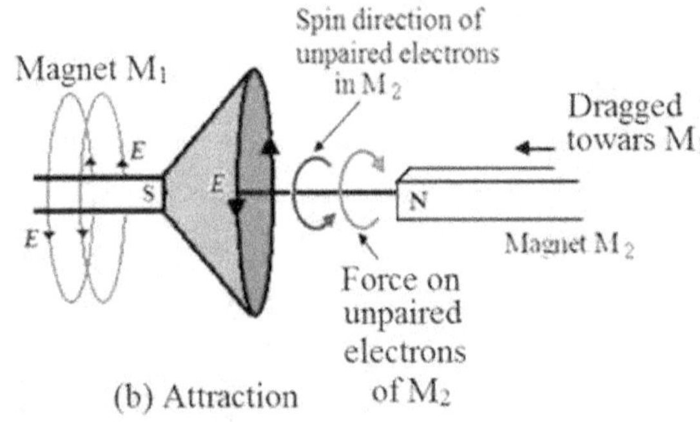

(b) Attraction

Figure 4.6.1. Attraction between poles of bar magnets.

4.6 Attraction and Repulsion between Magnetic Poles

The attraction and repulsion between magnetic poles are the reason why this subject began as a separate subject as magnetism. It is the attraction and repulsion that causes the magnetic properties to converge at both ends of the bar magnet, giving rise to the recognition of magnetic poles at both ends. But these two poles cannot be separated and they always exist in the form of dipoles. So much research has been done to see if they existed or exist independently anywhere in the universe or if they could be created in the laboratory but without success. Accordingly, many theoretical models have been proposed that predict the existence of magnetic monopoles, but they have not been found anywhere in the universe and could not be produced in the laboratory. Therefore, it may be very important to investigate if something else is responsible for the attraction and repulsion between the two ends of a magnetic bar. If some other factor is responsible, we will surely be numbed by the confusion of the concept of magnetic poles. We have some idea how bar magnets create a circular electric field around their poles. Figure 4.6.1 (a) shows the circular cone of S pole of a bar magnet. Both the effective spin current in that bar magnet and the electric field it creates are opposite in direction. An electron is placed above the axis of the bar magnet i.e. above the axis of the circular cone whose spin velocity is in the direction of the electric field. So the electric field of the bar magnet will try to decelerate the spin velocity of this electron so this electron will be pulled towards the strong region of the electric field of the bar magnet i.e. towards the bar magnet. This is what accounts for the attraction of two opposite poles of bar magnets. Two bar magnets A and B where S pole of A and N pole of B are close to each other as shown in figure 4.6.1 (b). In both bar magnets, the spin motion of the unpaired electrons which creates their electric field is in the same direction. The electric field of bar magnet A is in a position to decelerate the electrons in bar magnet B so the unpaired electrons in B will be attracted to the strong electric field i.e., magnet B will be attracted to magnet A. Figure 4.6.2 (a) shows the circular cone of S pole of a bar magnet. An electron is placed above the axis of the bar magnet i.e., above the axis of the circular cone whose spin velocity is opposite to the electric field. So the electric field of the bar magnet will try to accelerate the spin velocity of this electron so this electron will be pushed away from the bar magnet in the weak region of the electric field of the bar magnet. This is what accounts for the repulsion between the two identical poles of bar magnets. Two bar magnets A and B where S pole of A and S pole of B are close to each other as shown in figure 4.6.2 (b). In both bar magnets the spin motion of the unpaired electrons which creates their electric field is opposite to each other. The electric field of bar magnet A is in a position to accelerate the unpaired electrons in bar magnet B so the unpaired electrons in B will be pushed towards the weaker electric field of magnet A i.e., the B magnet will be pushed away from the A magnet. The same reaction magnet B will do with magnet A so that it will push magnet A away from magnet B. Due to the force applied by these two magnets on each other, a mutual repulsion will appear between them. Of course, to understand attraction or repulsion in magnets, it is not necessary to assume the concept of magnetic poles.

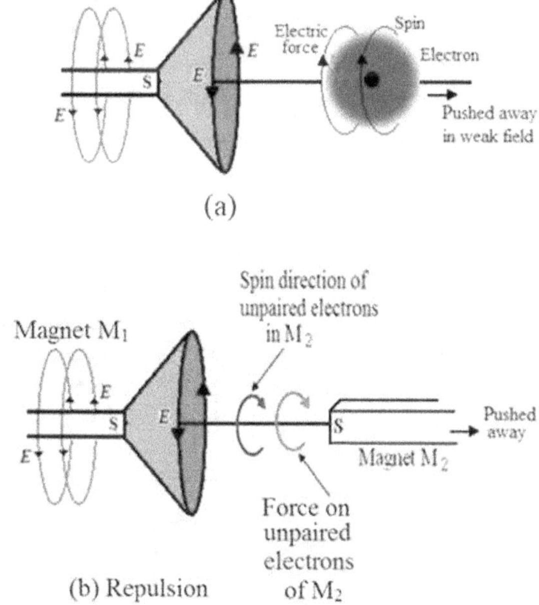

Figure 4.6.2. Repulsion between poles of bar magnets.

4.7. Magnetization in Magnetic Materials

Magnetic materials can be divided into two main categories namely magnetic materials and diamagnetic materials. We will primarily look at how magnetic materials acquire their magnetic properties. Prior to that a diamagnetic material is a material that is magnetized when placed in a magnetic field, however, the magnetization is in the opposite direction of the magnetic field. This means that generally diamagnetic materials are repelled by bar magnetics. These include strong diamagnetic materials such as superconductors and electron gas. A super conductor is one that has zero electrical resistance. At present, the resistance of the conductor is not zero, so super conductors currently have many advantages over conductors, but they do not exist at room temperature, so we are deprived of their advantages even today. Much research has been done around the world to see if superconductor materials can be produced at or near room temperature, but so far, no success has been achieved. Some materials exhibit superconductivity properties up to high temperatures of −135 °C, which are too low for practical use. In Figure 4.7.1 a superconductor is placed above the S pole of a bar magnet and is seen to be fixed in air at some distance from the pole. We know that unpaired electrons in a bar magnet result in an effective spin current which is circular as shown in figure. That effective current creates a circular electric field whose direction is opposite to that of the effective current. That electric field will form a circular cone near the S pole whose field strength decreases as it moves away from the pole. Placing a superconductor in that electric field will apply a force on its charges, the electrons. Those electrons will start moving in the opposite direction to the electric field due to the negative charge on them, because the electric field is circular i.e., the electric field is in the action of accelerating the electrons, they will be pushed away from the S pole which means the superconductor will be pushed away from the S pole i.e., up. Now the formation of an electric current in the superconductor will create a circular electric field around it which will act to accelerate the spin velocity of the unpaired electrons in the bar magnet. Hence a mutual repulsive force will be created between the bar magnet and the superconductor which is proportional to the effective current of the bar magnet. A superconductor will be stable in air at a position where this repulsive force and the gravitational force due to the Earth's gravity above the superconductor balance. This is called levitation of superconductor. At present this effect cannot occur for ordinary conductors.

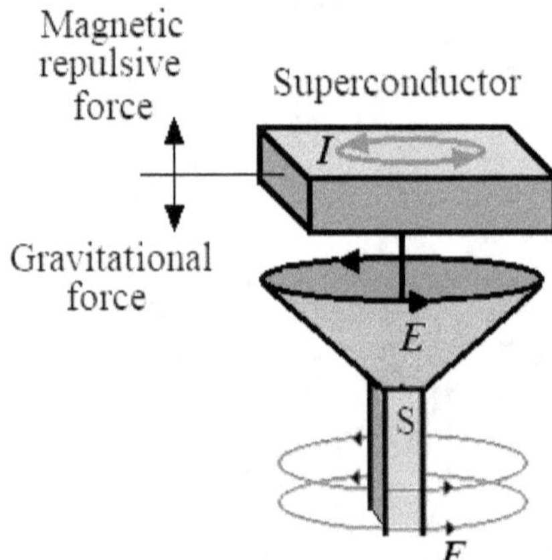

Figure 4.7.1. A superconductor floating in a magnetic field.

It is proposed in the theory proposed by L. D. Landau that the electron gas also acts as a diamagnetic material. But we also recognize that an electron has spin motion which is its intrinsic property, meaning that it is always in spin motion, whether it is inside an atom or outside an object. So, it is paradoxical how the electron gas will act as a diamagnetic material when it will always manifest itself as a tiny magnet. As we have seen in the Spin Atomic Model chapter, a diamagnetic material can only manifest itself when the electron must give up its spin property after exiting the element. A magnetic material usually has unpaired electrons in its elements to have magnetic properties. In paired electrons, since the spin motion of both electrons is opposite to each other, their net magnetic moment is zero. Unpaired electrons are responsible for creating magnetic properties. Even though the atoms in a substance have unpaired electrons, that substance does not necessarily have a net magnetic field or a net magnetic moment. Because the spin motion of all the electrons in it can be randomly oriented. In such cases their spins can be aligned in the same direction by applying an external magnetic field as shown in Figure 4.7.2 and a net magnetic field can be created around the material. This is called magnetization. A substance B, whose atoms have unpaired electrons, is placed in the magnetic field of a bar magnet A as shown in Figure 4.7.2. The circular electric field of the bar magnet will link the electrons in the material B and apply a force. So the net effect on the paired electrons will be zero but the spins of the unpaired electrons will align and the net spin current will be non-zero and hence an effective circular electric field will be created around it as shown in Fig. 4.2.2 (b). This is called magnetization. Now even if the magnet A is removed, the material B will still have the same circular electric field. This process will be understood in detail as follows.

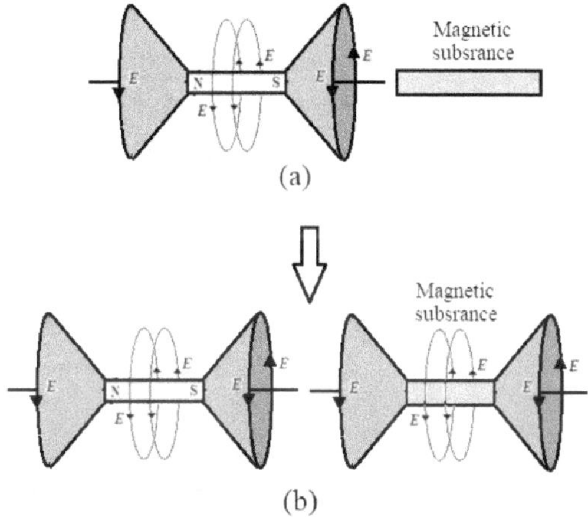

Figure 4.7.2. Magnetization of a magnetic substance.

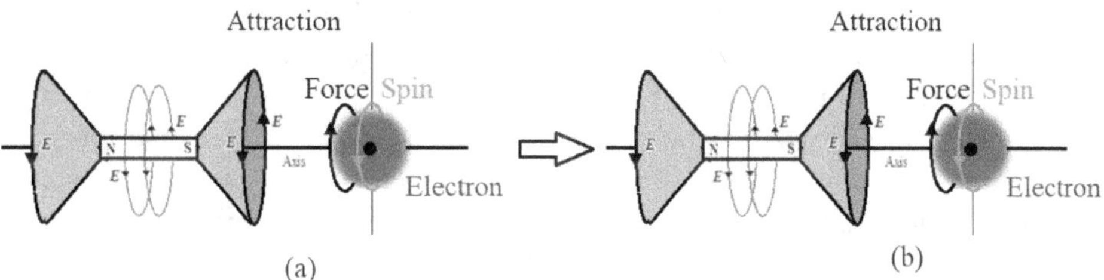

Figure 4.7.3. When the state of electron spin deceleration during magnetization of a magnetic material.

As shown in Figure 4.7.3, if the spin motion of an unpaired electron is in the same direction as the circular electric field created by the bar magnet, then the electric field will try to decelerate the electron. Therefore, the electron will be attracted to the magnet but will not be able to leave its parent atom and the spin motion it has acquired due to the parent atom will not be lost. Therefore, due to the electric field of that bar magn et, there will always be an attractive force on that electron towards the magnet, which will not allow to change the spin motion of that electron. Another possibility is to assume that when an unpaired electron in an item of matter is in spin motion opposite to the electric field of the bar magnet, as shown in Figure 4.7.3, the electric field will try to accelerate the spin motion of the electron. Then that electron will be pushed away from the bar magnet in the weak part of the electric field. But that electron cannot leave its parent atom so it will flip around its vertical axis, which is perpendicular to its spin axis, as shown in the figure.

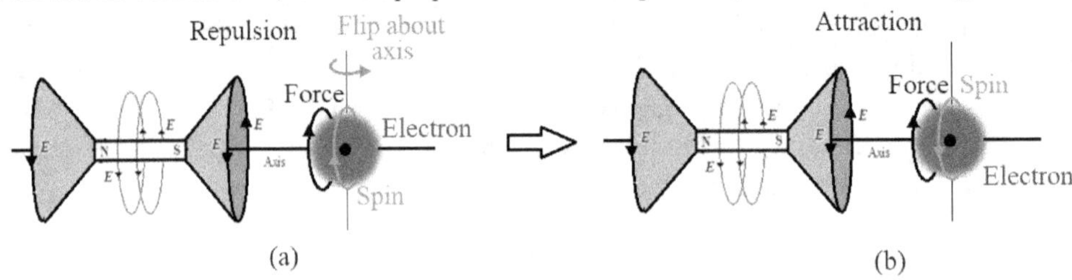

Figure 4.7.4. A state of electron spin acceleration during magnetization of a magnetic material.
Now the direction of its spin velocity will be in the direction of the electric field of the bar magnet, so it will be decelerated, so attraction will be created between it and the bar magnet, so there will be no further change in the spin orientation of the electron. Another thing that remains to be clarified is that if the unpaired electron in an atom is in the opposite direction to the bar magnet as shown in Fig. 4.7.4 (a). If its spin velocity is flipped by the electric field of the bar magnet, will it be in the same position? Or will come on the other side of the atom and gets spin aligned? That is yet to be clarified. But all the unpaired electrons will align and magnetize the material.

4.8 Electromagnetic Induction

Electromagnetic induction was discovered independently by Michael Faraday in 1831 and Joseph Henry in 1832. In his first experiment on electromagnetic induction, Faraday wrapped two separate wires on opposite sides of the iron ring shown in Figure 4.8.1 and expected that when an electric current was pa ssed through one wire, a wave would travel through the ring and would produce effect in the right -hand wire. He connected a galvanometer to one wire and when a power supply was connected to the other wire, he found that the galvanometer gave a momentary deflection when the power supply is connected to the other wire. This means that when the current started to flow through the primary wire when it was connected to the battery, a transient current was observed in the secondary wire and then it became zero. Again, when the connection of the battery was disconnected from the primary wire, the galvanometer also showed momentary deflection i.e., this time also transient current flowed in the secondary wire. Over the next two months, Faraday discovered several forms of electromagnetic induction and realized that a steady current flowed through a loop of wire as the magnetic field steadily increased or decreased. Or if the magnetic field linked to that wire keeps on increasing or decreasing, alternating current flows in that wire loop. He formulated a rule on this, namely,

'The electromagnetic force around a closed path is equal to the negative of the time rate of the magnetic flux enclosed by the path.'

Its equation is as follows,

$$emf = -\frac{d\phi}{dt}$$

where f is the magnetic flux.

Based on this, mechanical power is currently converted into electrical power, which includes wind turbines, thermal power stations, hydroelectric power stations, etc.

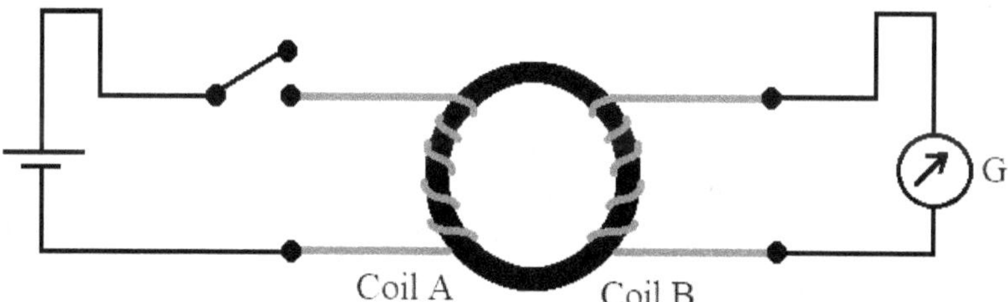

Figure 4.8.1. Faraday's Electromagnetic Induction Experiment.

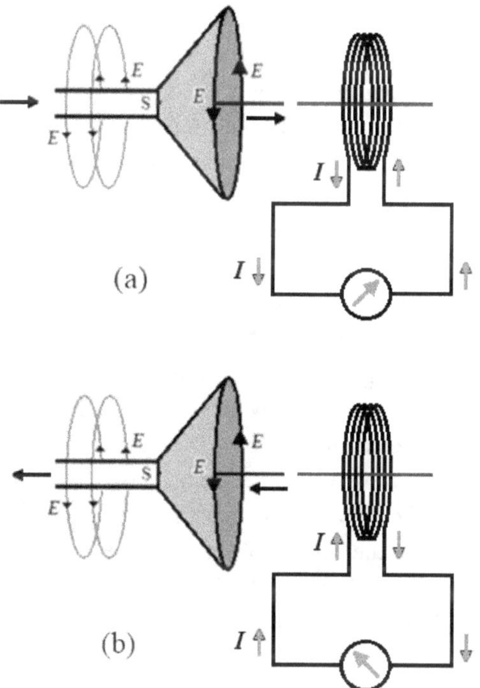

Figure 4.8.2. Explanation of electromagnetic induction by the circular electric field of a bar magnet.
One thing to mention here is that the magnetic induction B linked to that wire loop is known from which the magnetic flux f can be derived. Also, the electrons in the wire, which cause current to flow in the wire, we know the charge on them. They are moving so there is a force on them and obviously it is going to be a magnetic force. Now the equation of this force is also known then, why this force cannot explain this

electromagnetic induction? It may also mean that the magnetic field is not fully understood. We have seen that a magnet creates a circular electric field so it is necessary to see how it can respond.

Either magnets are rotated around a number of coils or magnets are rotated around the coils to convert mechanical power into electrical power. Hence, the change in magnetic flux linked to the coil with time is considered to be proportional to the electrical power generated by the Faraday's law. We need to get an explanation of how an EMF can be generated in a coil using the proposed circular electric field of a magnet. Let us try to explain how an emf can be generated in a coil using the proposed circular electric field of a magnet.

A coil is placed near the S pole of a magnet in Figure 4.8.2. Any S or N pole of the magnet can be used in this. The effective spin current of the electrons in the magnet will create a circular electric field around the magnet which is linked to the coil thus applying a force on the electrons in the conductor of the coil which is opposite to the electric field. Hence, the electrons that will be displaced will cause a current flow in the coil towards the electric field which will be sensed by the deflection of the galvanometer. But here one thing is noticed that this current is recorded only when the magnet is brought close to the coil. When the magnet is stationary near the coil, the current in the coil is zero and the deflection of the galvanometer is also zero, which is unexpected. Since the electric field is permanently connected to the coil, there should be a constant current in the coil, but it does not appear to be. But if the same coil is made of superconductor instead of ordinary conductor, there is continuous current flow in it, so why the current becomes zero in ordinary conductor is yet to be understood. In fact, the resistance of a common conductor is not zero, so the conductor will resist the flow of current. Due to that resistance, as the electrons flow, they adjust their spin motion in such a way that the effect of the magnet's electric field becomes zero and they stop moving. So transient current will flow until their spin motion adjusts and then stop. True, this explanation is bold, but all alternatives must be explored. When the magnet is moved away from the coil, the held electrons will lose their spin and their momentum will cause the electrons to move in the opposite direction to the first case and become zero again. This time the transient current will flow in the opposite direction to the first transient current as seen by the deflection of the galvanometer. So if the electric field linked to the coil increases and decreases alternatively, alternating current will flow in the coil and this fact is used in turbines. Another thing to be mentioned here is that when the magnet is approaching the coil, the electric field of the coil created by the transient current flowing in the coil will oppose the magnet from approaching the coil. Also, when the magnet moves away from the coil the electric field created by the coil will oppose the magnet moving away from the coil. So, the emf produced in the coil opposes the magnet to produce the emf which is called Lenz's Law. The phase difference between the electric field linked to the coil and the emf produced in the coil can be understood from the figure below.

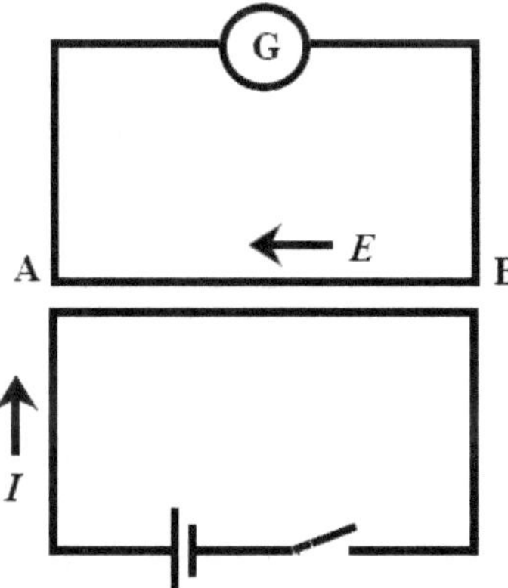

Figure 4.8.3. Circuit arrangement for induction of EMF.

We know that when an alternating current flowing through a primary circuit induces an EMF in a secondary circuit placed close to the first. We will actually see how that happens using asymmetric electric fields and asymmetric electric forces. Two circuits are placed close to each other as shown in figure 3.5.1.

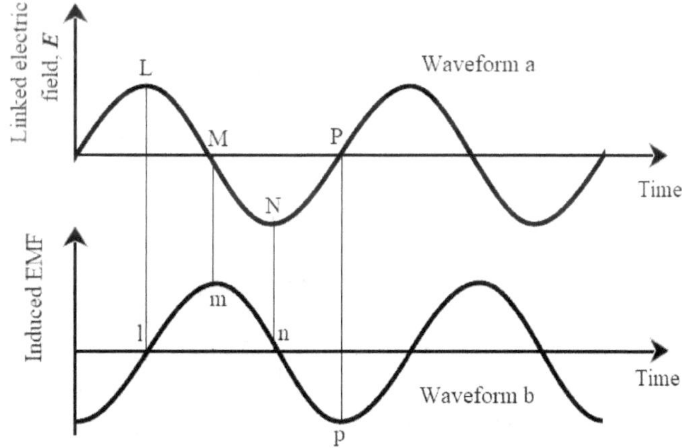

**Figure 4.8.4. Waveforms of the linked electric field and
the induced EMF in the secondary circuit**

Suppose, initially the conductor AB is not connected to the galvanometer. In a primary circuit a current I create an electric field parallel to the conductor and in the opposite direction to the current. This electric field polarizes the atoms in the conductor AB. This results in an excess of electrons at the B end and a deficiency of electrons at the A end and if these two points are connected to each other, the excess electrons from point B flow to point A which creates a transient current. Then again, the circuit comes to steady state and there is no more current in the secondary circuit. Actually, since there are free electrons in the conductor, one would expect a constant current but it seems that the electrons are not free. However, if the secondary circuit is made of superconductor, there will be a continuous current. Now if we break the current in the first circuit, the field connected to the second circuit becomes zero and the atoms which were initially polarized return to their normal state resulting in an excess of electrons at point A and a deficiency of electrons at point B. This results

69

in a transient current in the opposite direction as in the first case. Thus, it appears that if an alternating current flows in the first circuit, it can produce an equally induced EMF through the second circuits. Obviously, the induced EMF in the secondary circuit must be proportional to the magnitude of the field linked in parallel with the secondary circuit and to the rate varying with time. We connect an alternating source to the primary circuit to observe the phase difference between the induced EMF and the field linked to the secondary circuit. Then the field E linked to the secondary circuit is proportional to the alternating current in the primary circuit and let it represent the waveform as shown in Fig. 4.8.4(a). When the current in the primary circuit and hence the field connected to the second circuit reaches at point L, it remains momentarily constant and there is no induced current or induced EMF in the associated secondary circuit. In this condition the atoms in AB are polarized to their maximum level. When the electric field starts falling from point L to N, the polarization strength of atoms decreases and when the field reaches point N, it reverses. During this period the current in the second circuit flows in the direction AGB and finally becomes zero. Again, when the field starts to increase from point N to P, the current in the secondary flows in the direction BGA. The current in the secondary circuit and hence the induced EMF in it is represented by waveform (b) as shown in Figure 4.8.4. From these waveforms it appears that there should be a 90-degree phase difference between the coupled electric field and the induced EMF. While deriving the equation for the induced EMF, the following points should be considered.

1. The induced EMF in a circuit is directly proportional to the amount of the electric field linked in parallel to the conductor of the circuit.

2. The induced EMF in a circuit is directly proportional to the rate at which the linked electric field changes with time.

3. The induced EMF in a circuit lag behind the linked electric field an angle of 90 degree, which causes to introduce minus sign in the EMF equation.

Taking these into account, the equation for the induced EMF can be given as

$$emf = -k\frac{\partial}{\partial t}\oint \boldsymbol{E} \cdot d\boldsymbol{l}$$

where k is a proportionality constant that also adjusts for the effect of the cross-sectional area of the wire and E is the electric field connected to the conductor element dl.

The electric field E can vary in magnitude as well as along the length of the conductor. The integration is to be taken over the entire length of the conductor of the closed circuit in which the EMF is to be calculated.

In transformers, the EMF developed in the secondary coil is increased either by increasing the linked electric field or by increasing the length of the wire or by increasing both. To increase the linked electric field, ferrite materials are used at the core of the transformers. The electric field produced by the current in the primary circuit aliens the spin of the unpaired electrons in the ferrite material in such a way that the late electric field linked to the secondary coil gets increased. Actually, the field of the primary coil tries to revolve the electrons in its opposite direction but it seems that the electrons are not able to yield as their motions are strongly controlled by the rest of the charges in the atoms. In this interaction, the spin of the unpaired electrons gets aligned in such a way that they produce their net electric field in the same direction as that of the applied field that is why the net electric field linked to the secondary coil gets increased. If there are some free electrons in the core, then they revolve in opposite direction of the applied field. Therefore, the direction of the field produced by such electrons is in opposite to that of the applied field and thereby a reduction in the field linked to the secondary coil. These currents are called eddy currents. The eddy current losses are minimized by using a laminated core built up from thin plates of ferrite material tightly bounded with each other.

The superconductor possesses free electrons which can move with zero resistance, when a bar magnets electric field is applied to a superconducting material than the free electrons revolved in reverse direction of the applied electric field of the bar magnet. The electric field produced by these electrons in a direction opposite to the applied electric field as a result the net electric field is reduced, the repulsion between the super conductor and the bar magnet occurs as the applied electric field slowly decrease away from the bar magnet.

4.9 Summary

1. Only one field must exist in the universe and if that electric field is assumed then no other field can exist. There should have been a clear stand on this but no one did.

2. We assume that a current flowing in a straight line creates a circular magnetic field. The effect of that current is a magnetic field, but the simple question of how direct current can create a circular effect has never occurred to anyone.

3. The electric field in an electromagnetic wave applies a force on electrons. Force is mutual. Obviously, the electron's electric field must also apply a force on the charge associated with the wave's electric field. But the wave carries no charge, so the electron's electric field must apply a force on the wave's electric field. This means that the wave's electric field must be applying a force on the electron's electric field. This very important information should have come out but it didn't.

4. Human beings have imagined many things which do not exist in the universe and have thought about them. Similarly, if the asymmetric electric field had been imagined and thought about, then the real nature of the magnetic field could have been predicted and research could have been done.

5. From the above discussion, magnetic field is an effect of asymmetric electric field, so when the frequency of electromagnetic wave is increased, the magnetic force should increase. This thing needs to be proved immediately by experiments, which will radically change the way humans look at the universe. Also, this experiment will prove that the asymmetric electric field in the light wave is responsible for the photoelectric effect.

6. The concept that a single field must exist in the universe requires knowledge of the true nature of the magnetic field. Therefore, the question of what is magnetic field will also come to the fore. At least it would make sense that gravitational waves cannot exist.

Chapter - 05
Deflection of light rays in electric fields

5.1 Background

The general understanding is that light rays are an electromagnetic wave with no electric or magnetic charges. Of course, light rays contain photons and are just bundles of energy that we perceive as having no mass and no charge. Therefore, light rays will not deflect in electric or magnetic fields, because electric or magnetic fields can only apply force on electric or magnetic charges. In fact, when light rays are deflected in a gravitational field, it is naturally expected that electric and magnetic fields should also affect them. But due to the collapse of classical mechanics, these things could not come to light. Classical Mechanics being classical in name itself, it should have somehow come up as Fine Mechanics or Royal Mechanics but unfortunately it did not. Of course, even today it still has its place as a mechanics of common man. But what eclipsed this mechanics was that when this mechanics were emerging, the balance between the two things was not done. It means first to understand the philosophy of the universe in a rational way and to formulate that philosophy in the form of mechanics by verifying it through experiment. Philosophy involves the observation of the universe, its rational analysis and its interrelationship with other phenomena, and the search for its root cause. Mechanics aims to enrich philosophy by finding empirical evidence, formulating it into equations, and predicting the behaviour of the universe in detail. But in the last century it seems that philosophy has got a secondary place in the scientific world. As much and as quickly as possible, observation, hasty explanation and sensationalization has begun. So classical mechanics which initially emerged as Mechanics of the Universe but it did not get the proper addition of philosophy, was half-developed and thus could not explain much of what was happening in the universe. The reason for this is that from Aristotle, Ptolemy, Pluto until the last 200 years, the knowledge about the universe through logic was called philosophy. He did not want empirical evidence and its formulation through proper equations, so that philosophy would have been very unrealistic. When logic began to be empirically tested, the unreality of philosophy for the last 2000 years began to surface, and the result was that philosophy no longer made sense. So in the last two hundred years philosophy has been a barrage of nonsense and bombarded with new laws and corresponding equations from empirical evidence, so of course classical mechanics has also been bombarded with new laws. Naturally, the new rule meant that a new building and many more such buildings were created. They were independent of each other, that is, the universe was fragmented and became even more difficult to understand. The concept that the universe must have a single foundation somewhere was lost. Also, this universe began to seem to be something different from the idea, and from that idea, a new sense began to emerge, and from that new mechanics emerged, and that is 'Quantum mechanics. Common people's thinking and logic did not have much importance in it. What quantum mechanics has to say is incomprehensible and beyond the comprehension of the common man. Therefore, the combination of philosophy and mechanics became impossible. But philosophy is the explanation given by reason and reason is the answer to why this is happening. There is little mention of what quantum mechanics says about the deflection of light rays in electric and magnetic fields. Classical mechanics does not support this, but in order to clarify this point, it is necessary to review some of the classical mechanics itself.

72

5.2 Electrical Forces through Field-Field Interactions

Forces that come into existence naturally must have a cause. They must arise from some need and that need must be the same for all. That means there must be a single reason behind the existence of gravitational force, electric force and magnetic force. Earlier we saw that magnetic force is an effect of asymmetric electric force. Similarly, the gravitational force also needs to be solved. It is necessary to see if it can be connected somewhere between electric field and electric force. For this, if we understand the basic reason why the electric force comes into existence, this universe can be explained a lot. If you want to see if a light ray can be deflected in an electric field, you need to understand electric force and electric field properly. Earlier we have seen the generalization of electric force.

$$q_1 \xleftrightarrow{\; r \;} q_2 \qquad F = \frac{1}{4\pi\varepsilon_0}\frac{q_1 q_2}{r^2}$$

(a) Force through Charge-Charge Interaction

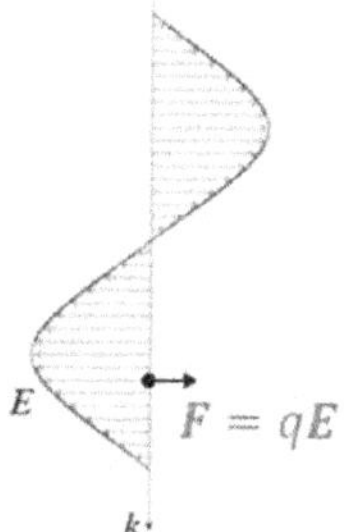

$$F = qE$$

(c) Force on the chrge q due to electric field E of EM wave. Force is mutual. Electric field of charge q will also apply force on the electric field E.

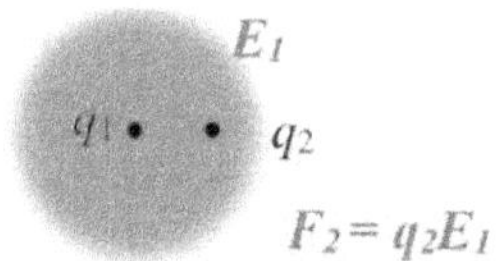

$$F_2 = q_2 E_1$$

(b) Fource through Field-Charge Interaction

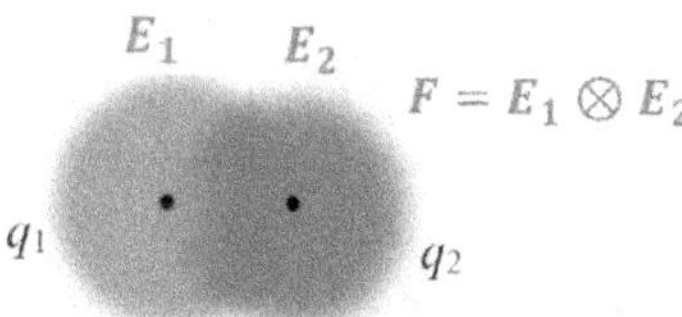

$$F = E_1 \otimes E_2$$

(d) Force through Field-Field Interaction

Figure 5.2.1. Generalization of electric force, (a) Coulomb force, (b) force applied by electric field on electric charge, (c) force applied on charge by electric field in wave, but force is mutual so charge will also apply force on electric field of the wave, and (d) both charges apply a force on each other through their electric fields.

As shown in Figure 5.3.1, Coulomb first formulated the law of force between two electric charges that the force between two electrical charges is directly proportional to the product of the charges and inversely proportional to the square of the distance between them. Where like charges repel and unlike charges attract. This force is mutual, i.e. the first charge applies a force on the second charge and at the same time the second charge also applies a force on the first charge. The concept of electric field was developed to understand how these two charges apply a force on each other when there is nothing or a vacuum in between. It states that a first charge creates an electric field around itself in which a second charge is placed and that electric field applies a force on the charge. Similarly, the second charge also creates an electric field around itself and thereby applies a force on the first charge. Although the concept of electric field is assumed to understand the interaction between two charges, its reality is realized by the transmission of electromagnetic waves. From this you can say that an electric field applies a force on electric charges, whose equation is $F = qE$. After this no generalization of electric force is seen. But one thing we have to consider is that when an electromagnetic wave propagates and interacts with a charged particle, the electric and magnetic fields in the wave apply a force on the charge. But the force is mutual which means that the charge must also apply force to the electric and magnetic fields. But with that electromagnetic wave there is no charge and force must exist. This means that the electric field of that charged particle must be applying a force on the fields of the wave. Assuming the magnetic force is the effect of an asymmetric electric field, the electric field of the charge must apply a force to the electric field of the wave. If this is true, then the wave's electric field must be applying a force on the charge's field rather than applying a force on the charge. That is, the electric force must arise from the interaction of two electric fields, which can be called force through field-field interaction. This requires a further generalization of the electric force. When the electric field of the charged particle applies a force on the electric field of the wave, the effect must be seen. A light wave is also an electromagnetic wave, so light rays can be deflected in an electric field. We are going to discuss this in detail from the point of view of obtaining practical evidence.

5.3 Deflection of light rays in an electric field

The first question is whether a charged particle can be deflected in a uniform electric field. It is impossible and can be explained from Figure 5.3.1. It consists of two electrodes which create a narrow but uniform electric field which is created by a layer of positive charges on electrode A and a layer of negative charges on electrode B. A light ray passing through this field will exert a force on both positive and negative charges in opposite directions as shown in the figure 5.3.1(b) and (c). In this the half wavelength of the light wave must be greater than the width of the electric field. A force will be applied on the positive charges in the direction of the electric field in the light ray and at the same time a force will be applied on the negative charges in the opposite direction. Since negative and positive charges are equal, both the forces will be equal but opposite in direction. Also, positive and negative charges will exert forces on the light ray due to their fields, but since both forces are equal and opposite in direction, a total zero force will be applied on the light ray, so the light ray will travel in a straight line without any deflection. This means that a field of only one type i.e., positively or negatively charged particles, means nonuniform field will be required to deflect light rays.

An electron can be thought of as a charged particle as shown in Figure 5.3.2 in which a light ray is shown passing along the electron. The moment the electric field of the electron and the electric field of the light ray are in opposite directions, there will be repulsion between them. Suppose the electron is fixed in space and does not leave its position, then the light ray will be deflected to the right as shown in the figure. Its deflection will depend on how close it is passing to the electron. The closer it goes, the greater the deflection.

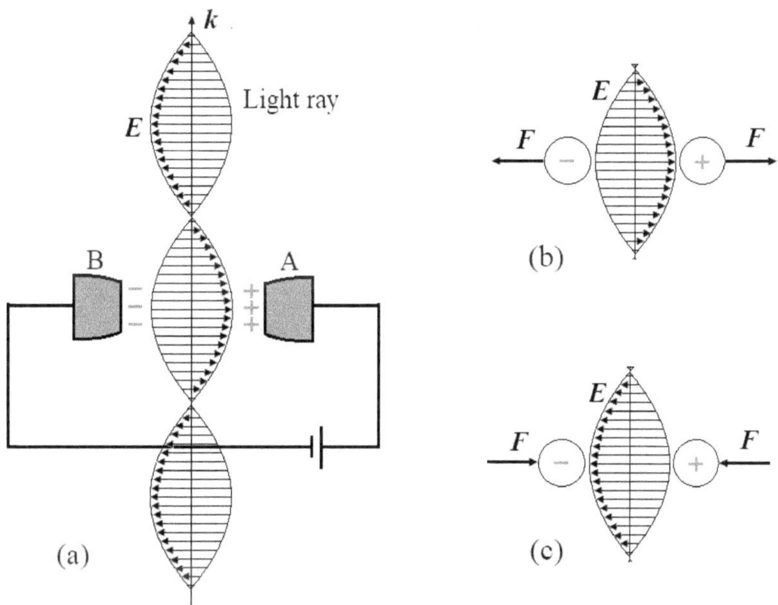

Figure 5.3.1. A light ray cannot be deflected in a uniform electric field.

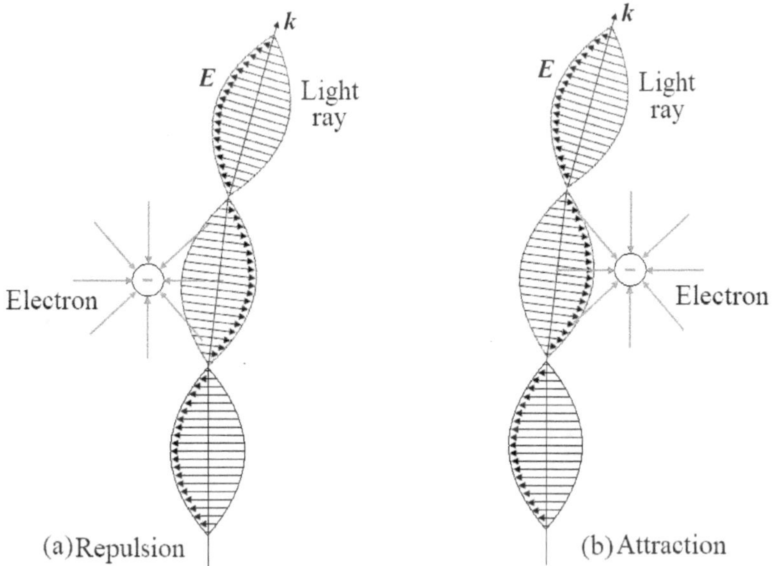

Figure 5.3.2. Deflection of light ray in field of an electron, (a) when attraction, (b) when repulsion.

A second light ray which is in phase with the first light ray but is traveling in a straight line to the left of the electron as shown in the figure 5.3.2(b) will also be deflected to the right. Because here is the attraction between electric field of the electron and electric field of the light ray. Both of those light rays can travel

75

further in a parallel line after being deflected. As illustrated in Figure 5.3.3, if a light beam passes closer to the electron than the first light beam, and they are of the same wavelength and in phase, the second light beam will be deflected more than the first light beam, and the two light beams will then will converge and interfere at a certain point. Depending on the path difference between the two, they will interfere constructively or destructively.

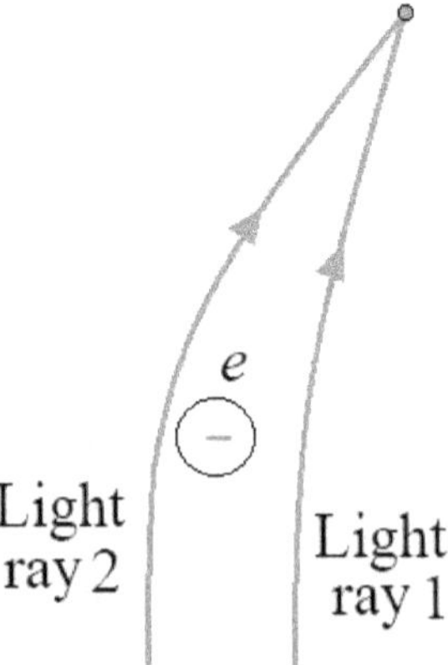

Figure 5.5.3. Deflection of light rays near an electro.

In all these examples all the light rays are plane polarized and polarized in the same direction. If the light rays from the laser light are polarized in the same direction and the laser light falls on the electron, then an interference pattern will appear on the screen as shown in Figure 5.3.4. When the electric field in the light beam, passing by the electron, is zero, the light beam will travel in a straight line and no interference pattern will be formed.

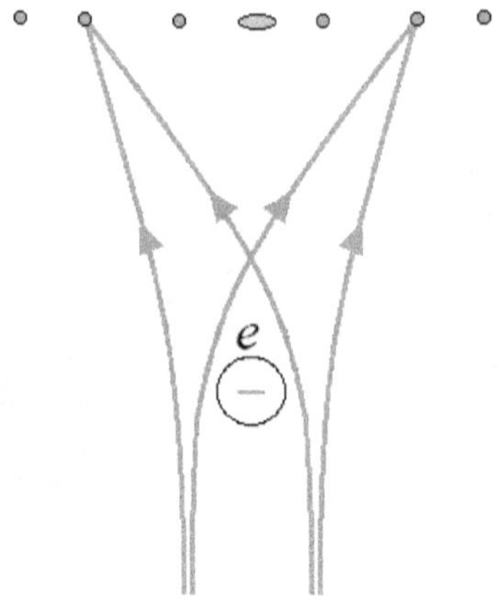

Figure 5.3.4. Light rays in a laser beam are forming interference pattern.

76

5.4 Electric force between two parallel light rays

We have seen earlier that the electric force arises from field-field interaction, so if two light rays are traveling parallel and close to each other, an electric force must exist between them. Of course, one has to understand first whether it will be attractive or repulsive. Of course, it will depend on how much phase difference there is between the two. If both waves are of the same wavelength, understanding will be even easier. We know that the waves in a laser beam are in phase so their dispersion is low, but the dispersion in ordinary monochromatic light waves is high because they are not in phase. If dispersion is to be reduced or eliminated, the light rays must be in phase and have the same plane of vibrations i.e., plane of polarization. If we want to understand when two light rays can attract, first we need to understand the attraction between electrons and protons. As shown in Figure 5.4.1 (a), an electron has a negative electric field around it and a proton has a positive electric field around it. Both fields are spherically symmetric and a plane P is drawn between them, where the fields of both particles are in the same direction. This means that when two light rays are close together and traveling in the same direction and are of similar wavelength and in phase as shown in Figure 5.4.1 (b), the electric fields between them will be in the same direction, so obviously there must be attraction. Perhaps this is why dispersion in a laser beam is low or negligible because the light rays in that beam are in phase.

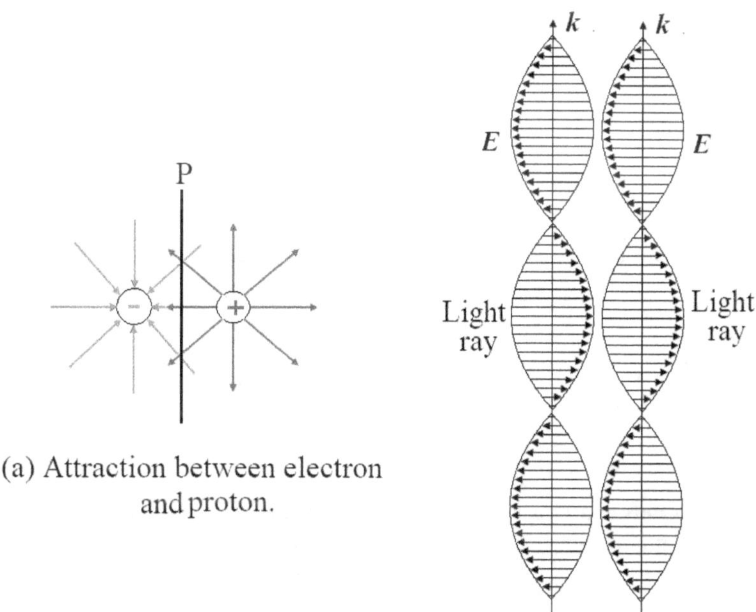

(a) Attraction between electron and proton.

(b) Two parallale light rays with in phase should have attraction.

Figure 5.4.1 (a) Attraction between an electron and a proton in terms of their fields, (b) Two parallel light rays in phase can attract each other.

To understand when two light beams can repel each other, one must first understand electron-electron or proton-proton repulsion. An electron has a negative spherically symmetric electric field around it as shown in Figure 5.4.2 (a) and when two such electrons are close to each other there will be repulsion between them. A plane P is drawn between their two fields, where the fields of the two particles are in opposite directions. The same will happen for two protons as shown in Figure 5.4.2(b). This means that when two light rays are close together and traveling in the same direction, they are of the same wavelength and when they are out of phase, as shown in Figure 5.4.2(c), the electric fields between the two will be in opposite directions, so obviously there must be repulsion between them. Perhaps this is why the dispersion seen in ordinary light beams is because the light rays are not in phase. Both these examples show that when light rays i.e., electromagnetic waves pass near each other, an electric force is definitely created between them. As we know that magnetic force is an effect of asymmetric electric field, it is not discussed here.

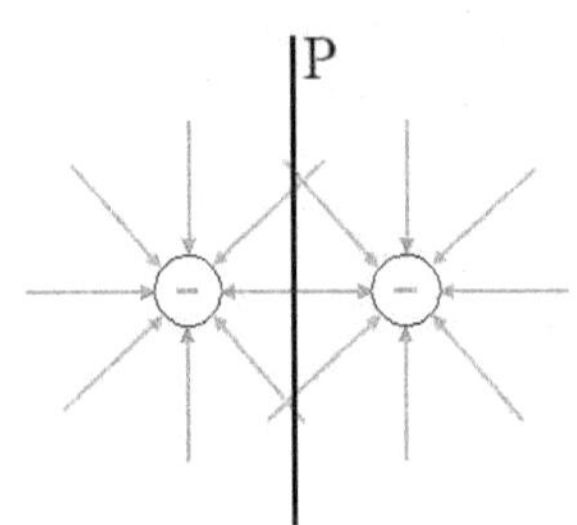

(a) Repulsion between electrns.

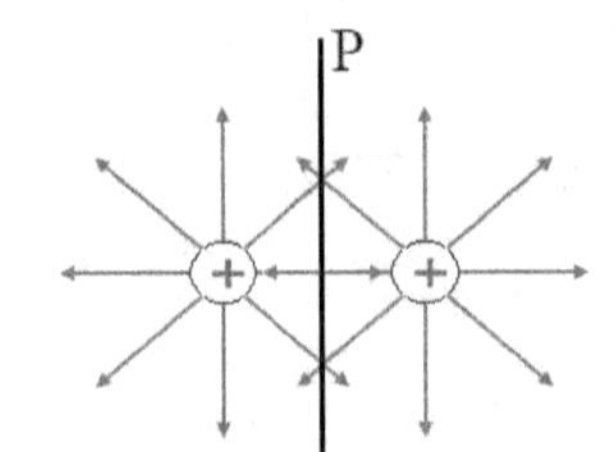

(b) Repulsion between protons.

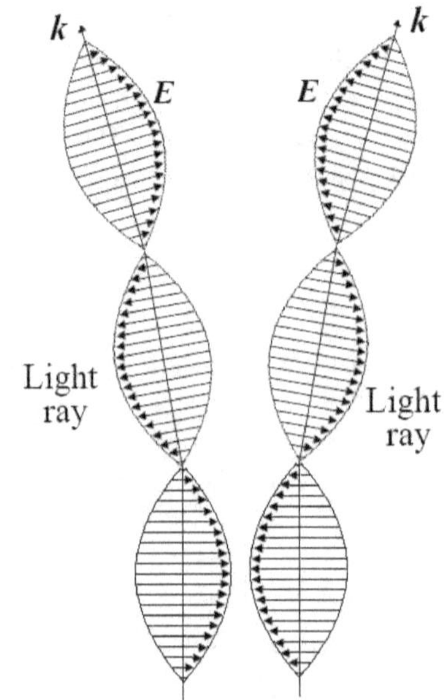

(b) Repulsion between two light rays
which are out of phase

Figure 5.4.2 (a) Repulsion between two electrons in terms of their fields, (b) Repulsion between two protons in terms of their fields, (c) Two parallel light rays which in out of phase can attract each other.

5.5 Summary

1. It is a universal truth that force is mutual. This is acceptable to all.

2. Electric force arises from field-field interaction.

3. The electric field of the electron will apply a force on the electric field in the light, so the light ray passing near the electron can be deflected.

4. A light ray can be deflected in a non-uniform electric field.

5. Two light rays that are parallel and in phase must be attracted. But to know how much and how it will be, it is necessary to study deeply, only then we can talk about it. If those two light rays are out of phase, they will have repulsion. Perhaps this is the reason why dispersion is seen in ordinary light beams.

Chapter - 06

Relative speed of light

6.1 Velocity of light in ether

Since light is an electromagnetic wave, it requires a medium for its transmission, and it was assumed that the speed of light through that medium must be constant. For this, ether was assumed as the medium and the big question was to find it. It would not be visible, and so to find the existence of such a thing one has to assume properties or attributes corresponding to it. An experiment has to be designed based on this property and its existence has to be verified. But if the proper attributes cannot be assumed, it is very difficult to discover the existence of such a commanding and invisible thing. It seems more likely that this is what happened with ether. Because even with tireless work, its existence cannot be found. The fact that any object can travel through the ether without friction means that the ether must be frictionless. Also, it must pass through all matter and occupy the universe and be a suitable choice as a rest frame. This means that the motion of every object must be calculated from the ether, and that motion of that object will be the true motion or real motion.

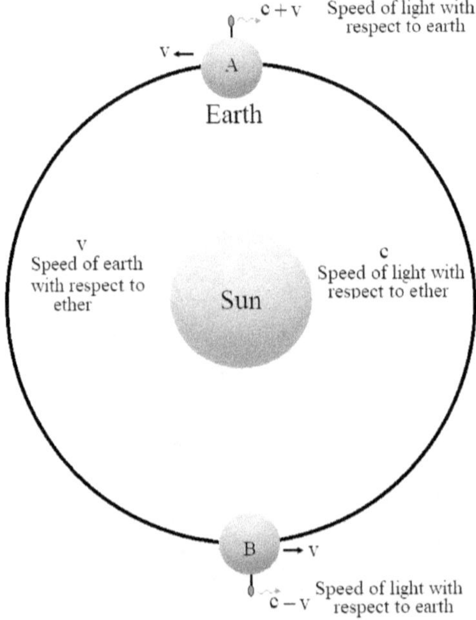

Figure 6.1.1. If velocity of earth through ether is v and speed of light through ether is c, then if the relative velocity of light from earth is measured then c+v should be filled at point A and c-v should be filled at point B.

For this, Michelson and Morley tried to measure the Earth's true speed from the ether, but they did not succeed. This made a huge difference in the way humans view matter and time. In this, if an object is at rest or in uniform motion in the ether, it remains as it is unless it is changed by the application of an external force. Hence, a frame of rest and a frame in uniform motion are called inertial frames of reference in which all physical laws apply equally. Motion is relative so one cannot tell who is at rest and who is in motion. And this seems to further hinder the discovery of the ether. In any case, the assumption that the speed of light must be constant relative to the ether seems reasonable. Of course, if experiments do not confirm this, there is nothing wrong with thinking that there is something else wrong with the hypothesis. There must be a correlation between the conclusions drawn by logic and the conclusions arrived at by experiment. Michelson and Morley, in designing the experiment, assumed that the speed of light through the ether must be fixed and that it is equal to c. Also, the earth revolves around the sun and it travels through the ether itself. Suppose the speed of the earth at point A through the ether is v as shown in the figure, after six months its speed at point B will be -v. If the light travels in the opposite direction to the speed of the earth at point A, then the speed of light measured from the surface of the earth should be c+v and if it is measured at point B, it should be c - v. The difference between the speeds measured from this would have given the true speed of the earth in the ether, but the speed of light appears to be the same in both places. Several times this experiment was tried very closely but every time no change in the speed of light was observed.

The Michelson–Morley experiment attempted to detect the presence of the luminiferous ether but became "the most famous failed experiment". In 1887, Albert A. Michelson and Edward W. Morley was the first to conduct the experiment to test the speed of light. The purpose of this experiment was to find the relative motion of all matter through the stationary ether using the motion of the Earth around the Sun.

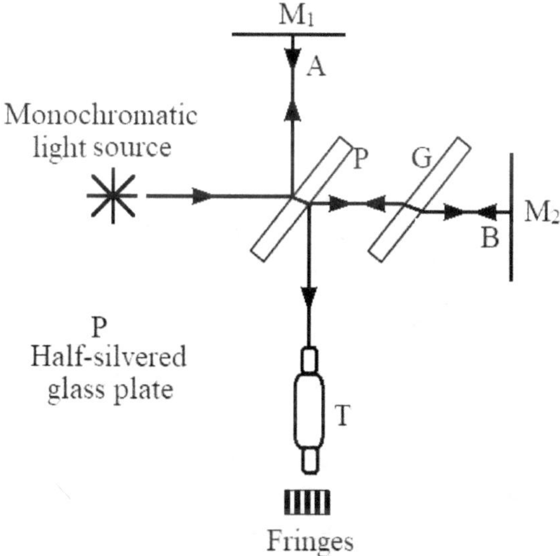

Figure 6.1.2. Design of the Michelson-Morley experiment.

Figure 6.1.2 shows the design of this experiment, in which a light beam from a monochromatic light source strikes a half-silvered glass plate P mounted at an angle of 45 degrees. This plate divides the light beam into two parts. The reflected part A travels vertically and is reflected at the normally flat mirror M1 and transmitted back to P and into the telescope. The transmitted part of the first light beam B is reflected from the plane mirror M2 and is reflected back at P and enters the telescope. The path difference between light ray A and light ray B will produce fringes and can be observed through the telescope. Ray A travels through glass plate P three times and ray B travels through glass plate P only once. A glass plate G is placed in the path of ray B so that the two rays travel same path. Although the distance traveled by both the rays is the same, the speed of the two rays will be different due to the device being placed on the surface of the earth and the earth moving around the sun. Because the speed of the light ray through the ether is constant and the earth is moving through the ether itself at a fixed speed, there will be a difference in the speed of the two rays. Once in that situation, measure the fringes and then rotate the device through a 90-degree angle to measure the fringes again and compare them. Given the speed of the Earth, later observations should show a 0.4 percent fringe shift from previous. But upon observation, it was found that only 0.01 percent of the fringes were displaced. No matter how carefully the experiment was performed and how often it was performed, the expected fringe shift was not observed.

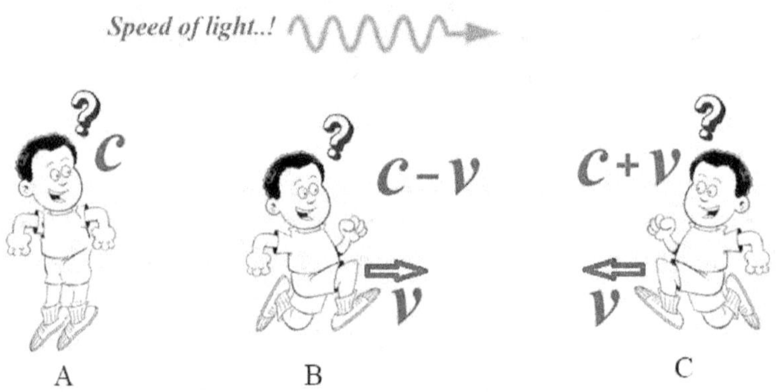

Figure 6.1.3. Relative speed of light, (a) observer stationary, (b) observer moving with speed v towards the light, (c) observer moving with speed v opposite to the light.

The pair were disappointed by the lack of success and did not discuss the results of the experiment much. But this conclusion later became very important and Michelson was awarded the Nobel Prize in 1907 for it. In fact, the conclusion of his experiment was that it was not possible to say with certainty whether the speed of light through the ether is constant or not. This experiment could have been successful only if the speed of light had been constant in the ether and thus the true speed of the earth in the ether could have been determined. But from the results of this experiment, Einstein assumed that the speed of light is absolutely constant and does not depend on the relative speed of the light source and the observer. Hence the very concept of ether was banished and all space was assumed to be a vacuum. This ignores the fact that speed is a relative thing. We assume that the speed of light is constant and is equal to c in vacuum. Suppose observer A measures the speed of light and records it as c. Also, observer B is coming in the opposite direction to the speed of light and has velocity v relative to observer A as shown in Figure 6.1.3, So his calculated speed of light must be c+v. Because both the numbers c and v are finite and known. Also, if the observer C is traveling in the same direction as the speed of light and his speed is v then he must fill in the measured speed of light c-v. But from the results of the Michelson-Morley experiment, Einstein hypothesized that the speed of light measured by all three observers equals c. This means that v has no meaning. Then how can c make sense because velocity is relative. In fact, both make sense so it should not make sense that the speed of light measured by all three observers should add up to c. Because speed is naturally relative, even if there is or isn't ether or what effect it has on the speed of light, there must be a difference between the speeds measured by these three observers. It seems that the light source that Michelson and Morley used in their experiment was fitted to the material of the experiment, and if the speed of light was constant relative to its source, the experiment would not have given the expected result. But later the scientist Signac proved through another experiment that the speed of light does not depend on the speed of its source. So, this matter has become more complicated. But it is true that the speed of any object depends on the relative speed of the observer, be it light or anything else. We are going to discuss some alternatives about this later, but due to the two hypotheses that Einstein put forward from the results of Michelson and Morley's experiments, we have gone astray. They are like this,

1. The laws of physics are the same in all inertial frames of reference.

2. The speed of light is constant; it does not depend on the relative speed of the source or the observer.

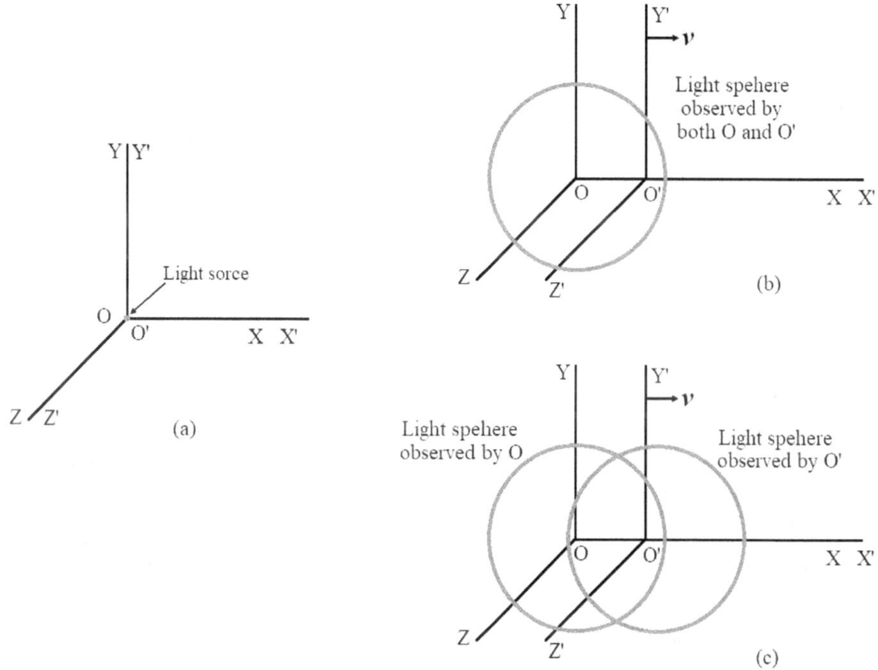

Figure 6.1.4. Observations of a sphere produced by the same light pulse by two observers, (a) O and O' overlap when the light source produces a pulse of light at t = t' = 0, (b) Both observers observe the sphere of light at time t = t', the expected sphere when observed must be the same according to a common man perception, (c) According to the Special Theory of Relativity, the two spheres observed by the two observers at time t = t' will be different, whose center is himself, which is beyond the imagination of a common man.

From this the transformation equations of coordinates were formulated called Einstein-Lorentz transformation equations. Two frames of reference O and O' are assumed as shown in Figure 6.1.4 where an interesting assumption is made that the X axes of both the frames always coincide with each other. Also frame O' is moving towards +X axis with speed v relative to O. The clocks of both the frames are different. At time t = t' = 0 the origins O and O' of both the frames are on each other when a light signal emitted from that origin propagates spherically. Now when the observer at O records the signal at time t, he perceives it to be spherical with its center at O. Also, when the same light signal is recorded by the observer at the location O' at time t', he also perceives the same signal as a sphere whose center is O'. At this moment the two observers are in different places, yet they both feel that we are at the center of this sphere of light. So, when a light is transmitted, does each observer experience it according to his convenience? This is never possible. Even if it seems so to the observer, the architecture will never change.

If we take such miracles for granted, they will not be without showing more miracles. Then we will delight in imagination rather than reality and create an imaginary picture of this universe rather than reality. The same thing happened with the Special Theory of Relativity. The funny thing about this is that no observer has yet been able to measure the direct speed of light, and no one is currently in a position to do so because it is so fast. Interference fringes option is used to measure it. When the direct speed of light is measured, which may soon be possible given the current rate of technological development, it will be known whether it depends on the relative speed between the source of light and the observer, and the foundations of the special theory of relativity will be shattered. As no alternative is currently available, the miraculous conclusions drawn by the Special Theory of Relativity are under discussion, namely Length Contraction, Time Dilation, Addition of Velocity, Mass-Energy Relation, etc. The first three of these are the result of the observer being in relative motion. If an object is in motion, the object's length will decrease in the direction of that motion. Also, if a clock is moving, it will appear to run slower than a relatively stationary clock, and if it is moving at the speed of light, it will appear to stop counting time. If an object is moving very fast, its speed can reach the maximum speed of light and cannot increase beyond that. And if another observer is moving in the opposite direction of the object's motion, he will find the object's measured speed to be no more than c so that $c + v = c$ and also $c + c = c$. Of course, all these assumptions are made when the object is in relative motion to the observer. If an observer goes along with the object, he will not see any change in the structure. An object is stationary to an observer and he measures its length and calculates L, and then when the object moves again to the same observer, when he measures it, he calculates L', and suppose that later the same object comes to rest near the observer and he again measures the length of the object. It will be L again. Now a person is your age and is with you and now he traveled away from you at a speed close to the speed of light so he will age slowly and if he comes back to you after a few years and stops, will he be your age now? because the length of the object is restored. There is no mention of what effect acceleration and deceleration in velocity may have on this. Another way to consider this special theory of relativity is that the experiment Michelson and Morley designed was intended to measure the constant or true speed of the Earth through the ether, and for that they assumed that the speed of light must be constant through the ether. The purpose of this experiment was not to measure energy, nor to determine the behavior of mass, nor was it measured at all, so how was it discovered that mass is equal to energy or that they convert to each other? It seems to have reached the point of writing equations upon equations. This is called reaching heaven on a thread. We think that energy is produced in nuclear fission and nuclear fusion according to the formula $E = mc2$ given by the Special Theory of Relativity. But as discussed earlier, if one neutron is equal to one proton and one electron, then in nuclear fission or nuclear fusion, the number of fundamental particles before and after the generation of energy is counted and if it is filled with the same number, the energy is generated only due to the rearrangement of those fundamental particles. This will come up. Because mass and charge do not exist. In the case of fundamental particles, the only thing that exists is the electric field. But in this situation, it will not be appropriate to comment more on this topic, it is necessary to think more about it. We are considering whether the speed of light is relative or not. There is no doubt that speed is relative. When trying to measure the speed of light, the source of light was fixed with the material of the experiment, whether it was the Michelson and Morley experiment or the Sagnac experiment. Another thing is that the speed of light is so high that it does not seem possible to measure it directly at present. Therefore, instead of measuring the direct speed of light, an attempt is made to find the change in speed by recording the change in wavelength. For this, an interferometer and the fringes formed in it are used. By seeing the current speed of technology development, in the near future when it will be possible to measure the speed of direct light, we will know what the truth is. By keeping the source of the light in this situation, or moving the observer, it is possible to observe the change in wavelength and determine the difference in the relative speed of the light. For that, two experiments can be done. But before that, it is also necessary to see what conclusions can be reached about whether the speed of light can have any effect on the speed of light from the experience of the common person in daily life and what conclusions have been reached by the scientists from such experience.

6.2 Starlight aberration

Aberration is a phenomenon which produces an apparent position of a celestial object from its true position. It is due to the relative velocity of the observer with respect to the celestial object which can be explained using figure 6.2.1(a).

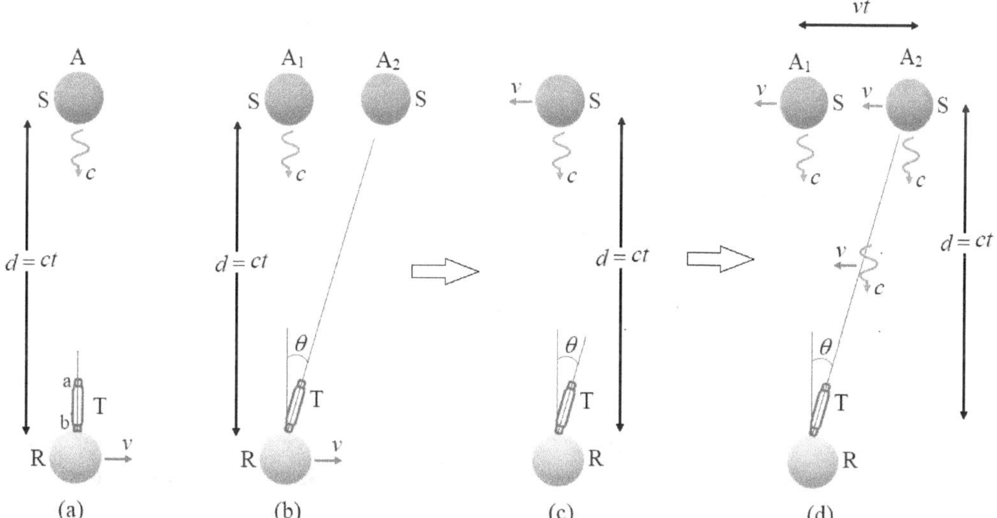

Figure 6.2.1. Starlight aberration, (a) and (b) observer in motion, (c) and (d) source in motion. A1 and A2 are true and apparent positions of source respectively.

Earth is moving with a velocity $v=3\times10^4$ m/s around the sun. An observer, if wants to locate a star above his head, let at position A by a telescope T, then he will not be able to locate its correct position. If he directs the telescope towards the correct position of the star, the light coming from the star when enters the telescope through point a and while going toward the point b, the telescope has been shifted towards the right side because of which the light will not reach to the point b of the telescope and hence the observer will not be able to see the star. If he tilts the telescope in the direction of motion, as shown in figure 6.2.1 (b), such that the light entered at position a will manage to reach at position b without any obstacle while travelling the telescope towards the right then he will be able to see the star. But now the observed position of the star will be at position A_2 instead of A_1 which is its correct position and A_2 is its apparent position. This phenomenon is confirmed experimentally and is called starlight aberration. Reverse of this event must happen is discussed next which carries more importance.

6.2.1 Starlight aberration in reverse

Velocity is relative. If an object A is moving with uniform velocity v relative to another object B in space means it is the rate of change of position of the object A with respect to object B or vice versa. It is impossible to predict that which object is rest and which is moving. It gives freedom to make choice of frame of reference in which either object A is rest or object B is rest. Therefore, starlight aberration should also occur when source is in motion and observer is rest as shown by figure 6.2.1(c). Only relative motion between them is needed. In this case also the observer will have to tilt the telescope in appropriate direction to locate the position of the light source or star as shown in figure 6.2.1(d). The light entering into the telescope now has two velocities, c and v, where c is the velocity of light and v is the velocity of the source. Thus, the velocity of the source is found to be imposed on the velocity of light. Without use of starlight aberration one can also prove this effect using the general motion of objects in space.

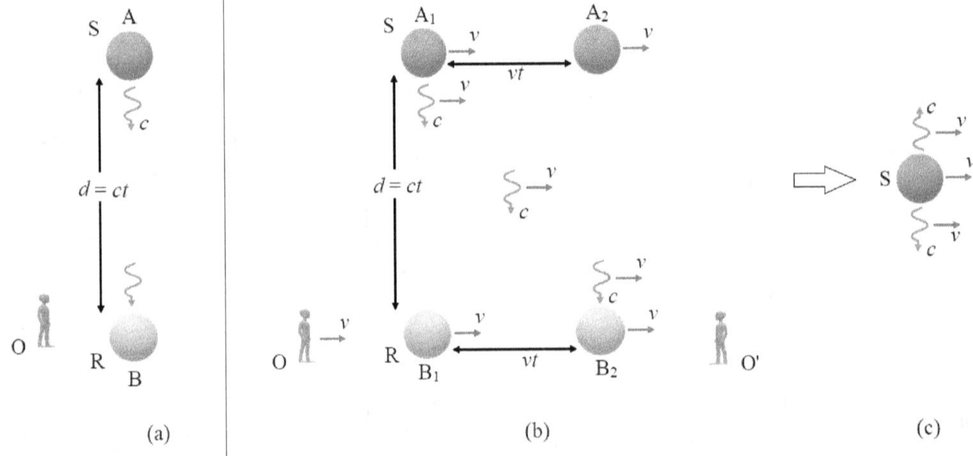

Figure 6.3.1. (a) Source, receiver and observer are at rest. (b) Source and receiver are moving with uniform velocity v to the right of the observer O'. (c) Velocity of source is imposed on velocity of light.

6.3 Effect of source velocity on light velocity: source moving perpendicular to light propagation

Consider figure 6.6.1(a) where a light source and a receiver are separated by a distance d. Source emits light pulse which receives by the receiver in time t, therefore, d = ct. An observer O, which is at rest, observers this event. Suppose there is another observer O' with respect to him the source S, the receiver R and first observer O are all moving with uniform velocity v to the right as shown in figure 6.3.1(b). The observer O' also observes the same event and finds that the receiver R receives the light at the same time t. But as the source and the receiver are in motion, therefore, he finds changes in positions while at the time of receiving the light. He notices that at time t=0, the source was at A_1 receiver was at B_1 and the light was emitted. In time t, the source has moved to position A_2 and receiver at position B_2 where the light is received. Thus, the light has travelled along the diagonal with velocity v to the right and with velocity c in downward direction. Thus, he finds that the velocity of the source is imposed on the velocity of light. It is expressed in terms of figure 6.3.1(c). One may think that how the observer O' have detected all the positions. For that one can use other technique to note this event. Suppose four observers having clocks synchronized are placed at positions A_1, A_2, B_1, and B_2 and note the event and then conclude together. They will conclude the same.

6.3.1 Experience in daily life

At first one may surprise that how it is possible. Consider another example of solid objects such as a canon fires a ball with velocity w to hit a target T as shown in figure 6.3.2(a). If observer is at rest with respect to the canon observes this event, then his observations will be similar to that of in figure 3(a). Now suppose the canon and the target are moving to right with velocity v with respect to another observer O' then his observations will be similar to that of observations made in a figure 6.3.1(b). He concludes that the ball fired by the canon has a two velocities w and v, where w is the velocity of the ball due to firing of canon and v is the velocity of the canon. These types of events we observe in daily life. Surprisingly this is also happening for light.

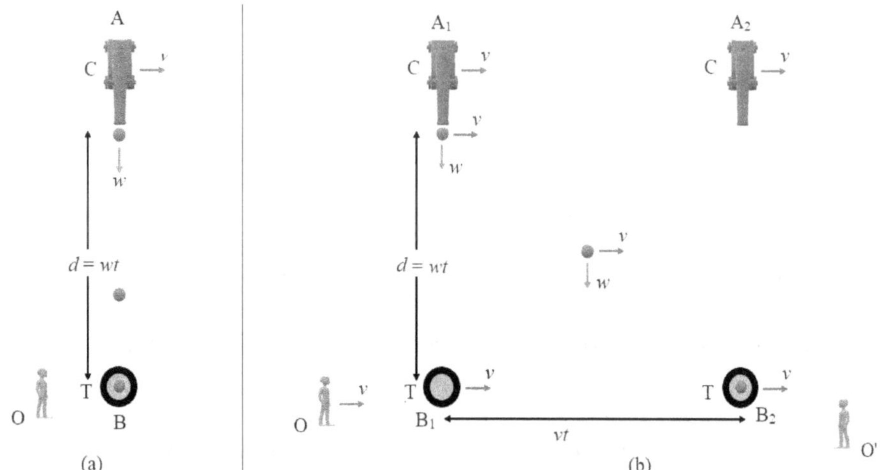

Figure 6.3.2. (a) Canon, target and observer are at rest. (b) Canon and target are moving with uniform velocity v to the right of the observer O'.

6.3.2 Experimental verification of light aberration when source in motion

In starlight aberration, observer on earth surface moves with orbital velocity 30 km/s around the sun and observes the aberration of light. Such types of aberration should also be observed if the light source is in motion instead of the observer. For that the source emitting light is to be moved with high velocity to get significant result. Fortunately, laser sources are available for production of sharp beam of light but to move them in high velocity may become a gruelling task.

Let us put a laser source on a rotating arm. The arm may be rotated with required speed to get high tangential velocity where the laser beam emerges in perpendicular to it. The expected aberration in laser beam may be observed at suitable distance from the source. Such type of designing is illustrated in figure 6.3.3 which consists of a portable laser source producing sharp beam of light fixed on one end of the arm. The screen is fixed at distance d and a slit in front of the laser source. At first, keeping the laser source S at rest, spot x_1 is marked on the screen where the laser light falls on screen. Further, rotate the arm with suitable rpm w and mark point x_2 on the screen of the laser light. Now it may be difficult to locate exact laser point on the screen due to the diffracted and scattered laser light from the edges of the screen. A telescope may be used for observations. Due to the rotational velocity of the shaft, tangential velocity attained by the laser source would be $2\pi Rw$, where R is radius of the arm. For instance, R = 3m, w = 3000rpm, the tangential velocity comes to, v = 942m/s. As, c = 2.99792×10^8m/s, it gives c/v = 318250.5 indicating to get 1cm displacement of the laser spot on the screen, for this rpm, the screen should be placed at distance 3.182505 km from the source. This distance appears to be large, but the results obtained will have exalted importance. Instead of setting the screen up to such large distance, one may take multiple reflections of the due beam in two parallel plane mirrors separated by a fixed distance and obtain suitable travelled path before examination. Such types of alternatives may be used, on demand of situations, at the time of performance of the experiments. Results of this experiment will confirm that the velocity of laser source is imposed on velocity of laser light when both velocities are perpendicular to each other.

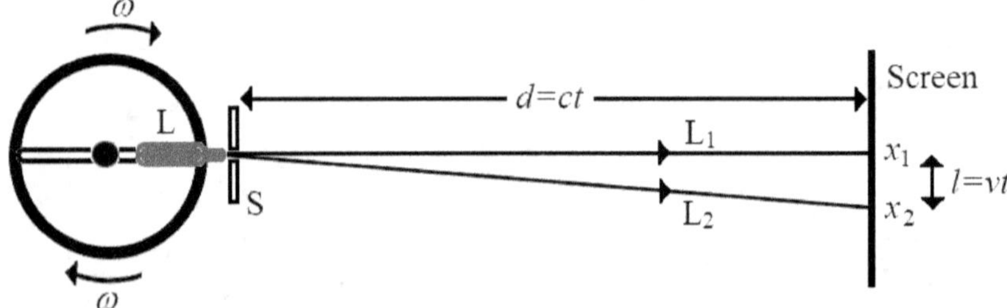

Figure 6.3.3. Laser light aberration experiment confirming velocity of laser source is imposed on velocity of laser light.

6.4 Effect of source velocity on light velocity: source velocity and light velocity are parallel

Now we investigate what effect could be produced on velocity of light when the source moves parallel to propagation of light. For that we consider figure 6.4.1 in which a source of light S placed at position A emits a light pulse in the direction of a receiver R placed at B which receives the light pulse. A rest observer observes this occurrence and confirms that the source emits light pulse at time zero and receives the receiver at time t with concluding distance between them is ct. Further, suppose the source S and the receiver R and the observer O too all are moving with uniform velocity v to the right side of the observer O' along a straight line, as illustrated in figure 6.4.1(b). Now source S emits a light pulse at time zero in the direction of the receiver R. As the source and observer are relatively rest with respect to each other, the receiver receives the pulse in time t as the distance between them is not altered. This phenomenon is now observed by the observer O' which is at rest and concludes that the light pulse is emitted by the source at position A1 in the direction of receiver at time zero. Meanwhile the receiver moved a distance vt to the right where it received the pulse at position B2 at time t. Thus, the observer O' concludes that the total distance travelled by the light pulse in time t is ct + vt with velocity c + v. Similarly, if the light is emitted in opposite direction of the source velocity by the source, then the velocity of the light pulse counted by the rest observer O' would be c - v. Thus, velocity of source is imposed on velocity of light when both are moving along same direction. It is illustrated by figure 6.4.1(c) including effect of figure 6.3.1(c).

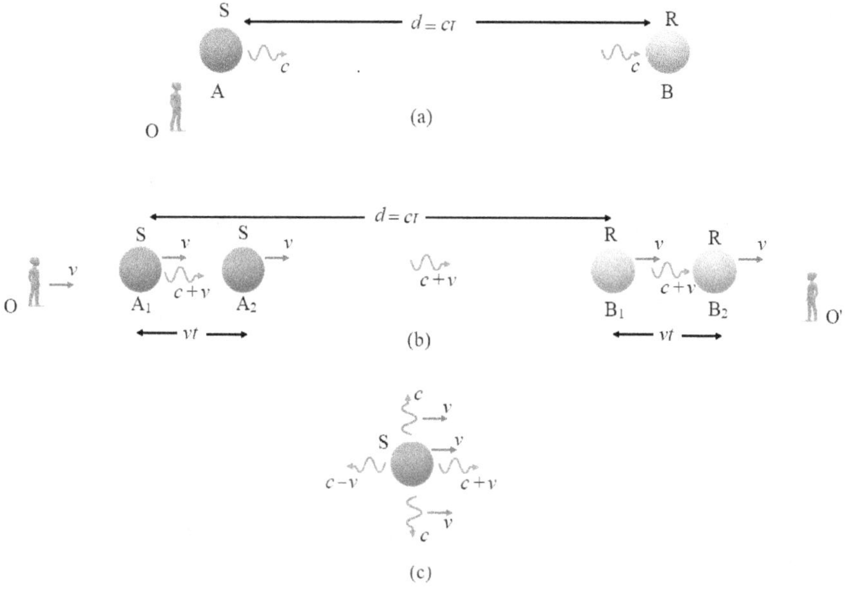

Figure 6.4.1. (a) Source, receiver and observer are at rest. (b) Source and receiver are moving with uniform velocity v to the right of the observer O'. (c) Velocity of source is imposed on velocity of light.

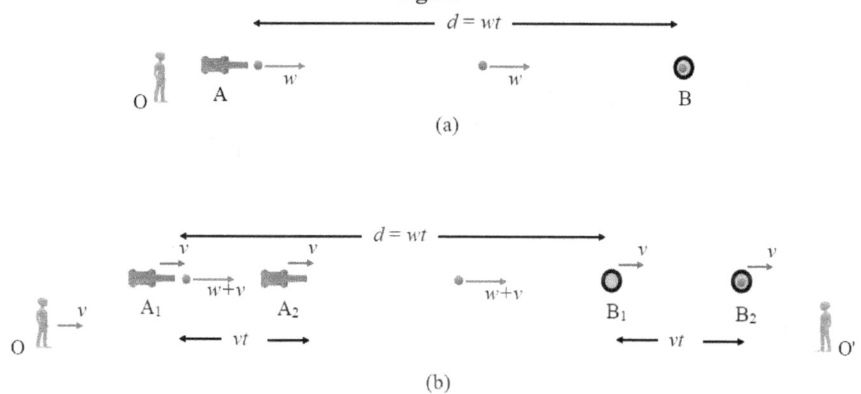

Figure 6.4.2. (a) Canon, target and observer are at rest. (b) Canon and target are moving with uniform velocity v to the right of the observer O'.

6.4.1 Experience in daily life

Such events occur frequently in daily life of common man which could be understood using canon ball firing as illustrated in figure 6.4.2. A canon fires a ball with velocity w to hit a target T placed at distance d at rest. An observer O which at rest observes this event such as the canon fires a ball at time zero which hits the target at time t with velocity w such that d = wt. Suppose the canon and the target both moves with uniform velocity v to right of another observer O' as shown in figure 6.4.2(b). This observer also observers the same event and concludes that, at time zero the canon is at position A1 and fires the ball which hits the target at time t. But meanwhile the target moved to distance B2 where the ball hits it. Thus, the observer O' concludes that the total distance travelled by the ball in time t is wt + vt with velocity w + v. If the canon moves in opposite direction of the target, the velocity of the ball observed by the observer would be w - v. Thus, the velocity of the canon is imposed on the velocity of the ball which a general experience of common man. This is also happening for light.

6.4.2 Experimental verification of velocity of source imposed on velocity of EM waves

Consider a system producing continuous electromagnetic wave of wavelength 0.2m. Surely, the wavelength measured by an observer, when both source and the observer are at rest or both are in uniform motion in space moving with same speed in same direction, is constant and is evidently 0.2m. If the source is made to move with velocity 3600km/h towards the observer, the changed wavelength observed by the observer using Doppler Effect equation

$$\lambda_o = \lambda_s \sqrt{\frac{1 - v/c}{1 + v/c}}$$

should be 0.199999333m. This wavelength is shorter than the previous wavelength. Now such two waves, of wavelength 0.2m and of 0.199999333m, are made to travel along same direction simultaneously. If these waves, at any instant, are in phase at point A then, because of the wavelength difference, they will be out of phase at distance nearby 30km, say at point B, from the initial point A. For further travel of 30km they will be again in phase.

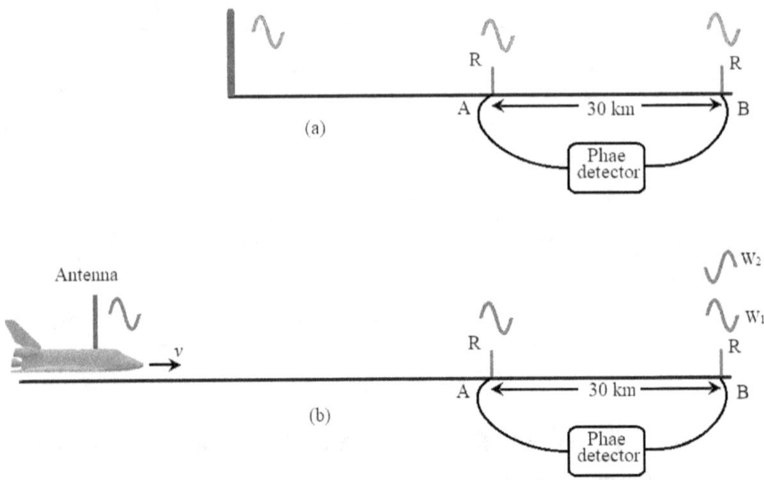

Figure 6.4.3. (a) Radiating antenna is at rest. The EM wave at positions A and B is in phase. (b) Radiating antenna moving with velocity 3600km/h towards the receivers where the EM wave at positions A and B should be either in phase (W1) or out of phase (W2).

For experimental verification, we set an antenna producing such continuous wave on ground surface. First, we mark a point A on the ground and draw a line passing through the antenna. We mark another point B on the line at distance 30km from point A and place two receivers at these two points to receive signal produced by the antenna. The received signals are fed to a phase detector to detect the phase difference between these two signals received at these two points as shown in figure 6.4.3(a). Now we allow the antenna to produce the EM waves of 0.2m wavelengths and adjust position of any one of the receivers to get the phase difference zero between the signals received by the receivers.

Now we make the antenna to travel along the line towards the receivers with velocity 3600km/h and check the phase difference between the signals received by them as shown in figure 6.4.3(b).

1. If the phase difference is of 180 degrees, then the velocity of the wave is constant and frequency and wavelength is changed according to the Doppler Effect. It means velocity of the source is not imposed on the velocity of wave.

2. If phase difference is zero then wavelength is not changed but velocity and frequency is changed. It means velocity of the source is imposed on the velocity of wave.

What does common man expect?

The presence of wave at point A or B means it produces electric field and magnetic field at these points. Production of fields means polarization of the vacuum. It will not depend on who is observing, whether the observer is at rest or in motion. In figure 6.4.3(a), suppose another receivers A' and B', separated by the same distance that of between A and B, are moving towards the antenna and observe the phase difference at the time when they coincide on A and B. The phase difference observed by both systems should be same as the field properties produced there should not depend on the state of observer. Since the receivers A and B are at rest and observe zero phase difference. But according to the Doppler Effect the phase difference observed by the receivers A' and B' should be of 180 degrees as they are moving towards the antenna with velocity 3600km/h because of which the wavelength observed is shortened. Common man expects that there should not be change in wavelength consequently velocity and frequency should be changed implying that the velocity of the antenna should be imposed on the velocity of wave. Since velocity is relative and no one can decide whether antenna is moving or the receivers are.

6.5 Measurement of change in wavelength using grating

Wavelength of monochromatic light can be determined by using grating. The greeting consists of vertical lines drawn in black ink on the glass as shown in Figure 6.5.1. A greeting commonly used in a physics laboratory can have up to 1500 lines per inch called the greeting constant N. Wavelength of light is given by using grating as follows.

$$\lambda_s = \frac{\sin\theta}{nN}$$

where, λ_s is the wavelength of the source,
θ, the angle of the order,
n, the order of the spectrum and
N is the grating constant.

The design of this experiment is shown in Figure 6.5.1. When the source and greeting of light are relatively stationary, let the wavelength of the light be λ and the frequency is f, then the speed of the light is $c = \lambda f$.

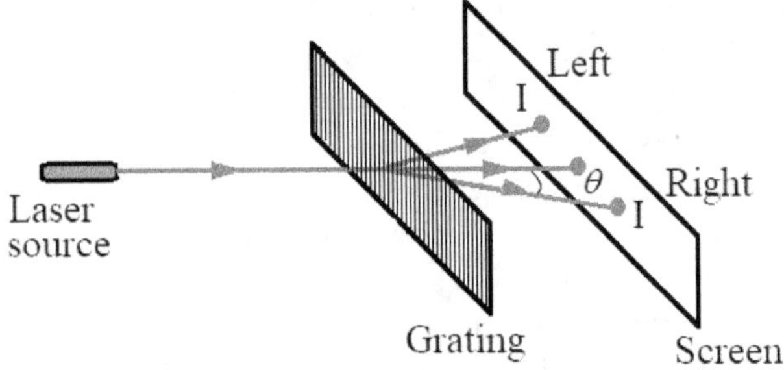

Figure 6.5.1. Grating experiment to calculate the wavelength of monochromatic light.

If in this experiment we move the light source towards the greeting or the greeting towards the light source at a speed v then the wavelength and frequency of the light will change. The expected change in wavelength can also be calculated from the following Doppler equation.

$$\lambda_o = \lambda_s \sqrt{\frac{1 - v/c}{1 + v/c}}$$

where, λ_o is the wavelenth observed by the observer.
λ_s is the wavelength of the source
v is the velocity of the laser source towards the garting,
and c is the speed of light.

If the light source is not moving towards the greeting with speed v and the first order angle q' is recorded and it is different from the first order angle q, then

$$\lambda_o = \frac{\sin\theta'}{nN}$$

where, λ_o is the wavelength of the moving light source,
θ, the angle of the order,
n, the order of the spectrum and
N is the grating constant.

From this grating equation the wavelength of the light can be calculated.

In this experiment, if we use a He-Ne laser light source with a wavelength of 632.8 nm and a grating constant of 15000 lines per inch, the first order angle is given by equation,

$$\theta = \mathrm{Sin}^{-1}(\lambda nN)$$

will be 21.674828 degrees.

If the light source is moving towards the grating at the required speed and no change is observed in the angle of any order of the grating, then it is assumed that there is no change in the wavelength of the light but there is a change in the frequency of the light according to the law of doppler effect and it should be

$$f_o = f_s \sqrt{\frac{1 + v/c}{1 - v/c}}$$

where, f_o is the frequency observed by the observer.

f_s is the frequency of the source

v is the velocity of the laser source towards the garting,

and c is the speed of light.

This will prove that there is no change in wavelength but a change in frequency and hence in the speed of light as its source is in motion which would be

$$c' = f_o \lambda_s$$

This would mean that the speed of light has changed due to the speed of the source and would be equal to

$$c' = c + v$$

whether it is that much can be investigated.

And if the light source moved towards the grating at the correct speed, and a change in the angle of any order of the grating is observed, then the wavelength of the light has changed.

And from that the changed wavelength l' can be deduced and also the changed frequency of light f' can be deduced according to the law of doppler effect and from that the speed of light can be deduced and it is back

$$c = f_s \lambda_s = f_o \lambda_o$$

to be obtained.

This means that even if the source of light is in motion, there is no change in the speed of light and this result can be considered to be in accordance with the Special Theory of Relativity.

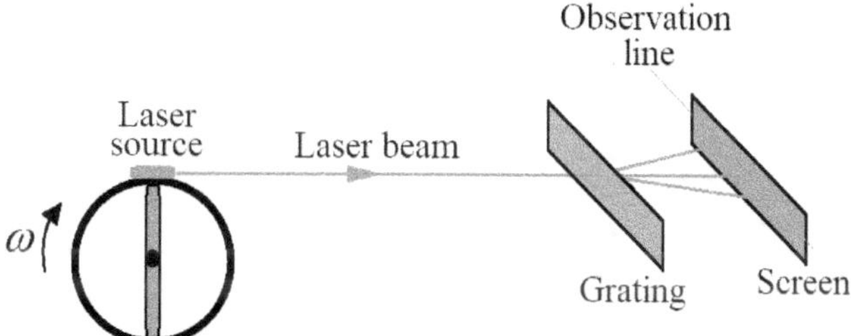

Figure 6.5.2. Grating experiment design with monochromatic light sources (laser sources) mounted on a rotating arm.

In order to get the right result through this experiment, the speed of the laser source towards the grating or the speed of the grating towards the laser source should be very high. How much it is needed will depend on how much difference you expect between the first or second order angles of the grating. To achieve such a high speed, perhaps the laser source can be mounted on top of the rotating arm as shown in Figure 6.8.2 to obtain the appropriate tangential speed and achieve the desired effect.

This experiment will prove whether the speed of light sources really affects the speed of light. Actually, the speed of the light source should have no effect on the wavelength of the light but should affect the frequency means the speed of the light. This experiment can prove what is true.

6.6 Summary

The following conclusions can be drawn from the above discussion.

1. Ether, if it exists, allows any object to continue its uniform motion without restriction.

2. A light source consists of charged particles and fields. Thus, along with the source, the field associated with the charged particles also maintains a uniform motion in the ether without any restriction.

3. Fields of charged particles in the atoms of the source are responsible for producing light which also contains fields. Therefore, the velocity of the charged particles is superimposed on the velocity of the field in light. In other words, the speed of the source is always superimposed on the speed of light.

4. If the ether could have prevented the uniform motion of the objects in it and tried to stabilize them, the speed of light with respect to the ether would have remained constant.

5. If the ether does not exist, then the vacuum is doing it.

Chapter - 07

Ether, Particles, Fields and Waves

7.1 Thoughts of the Ether by Scientists

The word ether comes from the ancient Greeks. It means the upper region of space where heaven is believed to exist. In the 4th century BC, the Greek philosopher Aristotle proposed the name ether for the fifth classical element proposed by his teacher Plato, such as water, fire, air, earth, and ether.

Even ancient Indian philosophy seems to have recognized the existence of Panch Mahabhutas. Mahabhutas are the elements from which the universe is created and the five Mahabhutas are earth, water, fire, air and sky. Of course, Akasha means Ether, the principle that pervades the entire universe. It is He who is considered as the Supreme Soul in Indian spirituality which is responsible for the existence of every particle. Until the beginning of the 15th century, the scientific world does not seem to have changed much about the perception of the universe. The existence and happenings of all things in the universe were deduced by logic and it was considered as philosophy. Therefore, such thinkers were considered superior as philosophers. Naturally, the confirmation of reasoning by demonstration was considered a secondary task. Therefore, centuries after centuries, there is not much change in human's view of the universe. But from the 15th century, some scientists, especially Copernicus, Galileo, tried to break the concept of the universe based on religion that had been going on for years by observing the universe from a scientific point of view. In the subsequent period, every phenomenon happening in the universe was analysed through logic and confirmed from a scientific point of view through experiments, and the knowledge obtained from it came to be called science.

Hertz discovered electromagnetic waves in 1880, and the question arose as to what medium should be used to generate and transmit these waves, such as air for sound. Therefore, Ether medium was assumed for the propagation of electromagnetic waves, it is also called luminiferous ether. Modern scientists such as Nikola Tesla believed that the ether was responsible for explaining basic natural phenomena such as gravity and light. Many scientists were convinced that the ether theory was correct and continued to explore it. The existence of ether was first tried to confirm by the results of experiments conducted by Michaelson and then by Michaelson and Morley in joint papers between 1885 and 1887. He postulated that if the ether is the medium responsible for the generation and propagation of an electromagnetic waves, then the velocity of the electromagnetic wave passing through the ether must be constant relative to the ether. He constructed an interferometer known as the Michelson-Morley interferometer, imagining that matter travels freely through the ether and that light rays pass through most matter. The purpose of this interferometer was to record the change in the speed of light as the Earth moves in one direction and then in the opposite direction as it orbits the Sun, as well as by rotating it through 90 degrees. From this, the speed of the earth compared to the ether was to be understood. Because as the Earth revolves around the Sun, its speed will be different through the ether and therefore the speed of the Michelson-Morley interferometer placed on the Earth will also be different during the year. Therefore, the speed of the light rays he recorded should also have been different. But his tireless observations did not detect any change in the speed of light rays, thus questioning the very existence of the ether. Because the conclusion of this experiment was that the speed of light is always constant and does not depend on the relative speed of the source of light and the observer. This finding was quite surprising as speed is always comparable but was not seen here. This conclusion contradicted directly the concept of speed. It was on this basis that when Einstein published his Special Theory of Relativity, the physics community began to accept that the ether did not exist. Now we are going to trace this once again.

7.2 Fundamental Particles and Forces

All the thinkers and scientists had an important question and that is where did matter come from? which is responsible for our existence. Matter is the first thing that comes to our mind as solid or solid matter, then when we think of liquid matter and after that we think of gaseous matter. Of course, the smallest particle of matter which we call atom is considered indivisible. Further research revealed that the atom is made up of three basic particles: electrons, protons and neutrons. Different atoms have different number of elementary particles, which gives the atom certain properties. These properties lead to the formation of different substances. The purpose of this paper is to find out what these fundamental particles, known as fundamental particles, are formed from, and do they have anything to do with the ether? Of course, a lot of research has been done on this and more is going on. Many subparticles have been discovered that are responsible for the creation of these fundamental particles, most prominently the Higgs bosons, also known as God particles. That is, it is considered to be the building block of the creation of this universe. Attempts to build it at CERN, where the Large Hydron Collider is the accelerator center, were unsuccessful. It is circular and its radius is 4000 meters. It costs 4.75 billion dollars and is contributed by 22 member countries including Belgium, Denmark, France, Federal Republic of Germany, Greece, Italy, Netherlands, Norway, Sweden, Switzerland, United Kingdom and Yugoslavia. CERN employs approximately 2,500 people from its member countries, but its facilities are available for use by research institutions around the world. In total, nearly 11,000 scientists of more than 100 different nationalities use CERN's infrastructure. Two protons were accelerated in opposite directions in this accelerator to produce Higgs bosons, or God particles, and collided when each had a kinetic energy of 7000 giga electron volts. The energy produced from their collision was expected to produce Higgs bosons, but this did not happen. Some scientists thought that once the Higgs bosons were created in the laboratory, they would devour all the matter in the universe and the universe would end. Many such interesting stories are associated with it. Seeing this vast expanse of research, one feels that it is not possible for a man to comprehend it in his entire life. Therefore, if one wants to understand this universe, one does not even know where to start and one becomes disappointed. It seems beyond the intellect of the common man, yet research is going on round the clock, brick upon brick is being built every day, floor upon floor is being raised. How our brick is right and how it fits next to this brick is always an effort and the building grows. No one seems to be able to establish the connection between the foundation of this building and the top floor, which is under construction. Because the building seems to be getting more and more complex, it is getting directionless. That interesting idea is based above, away from reality, and since this edifice rests on the brain of man, he cannot think differently from it, which is in keeping with reality.

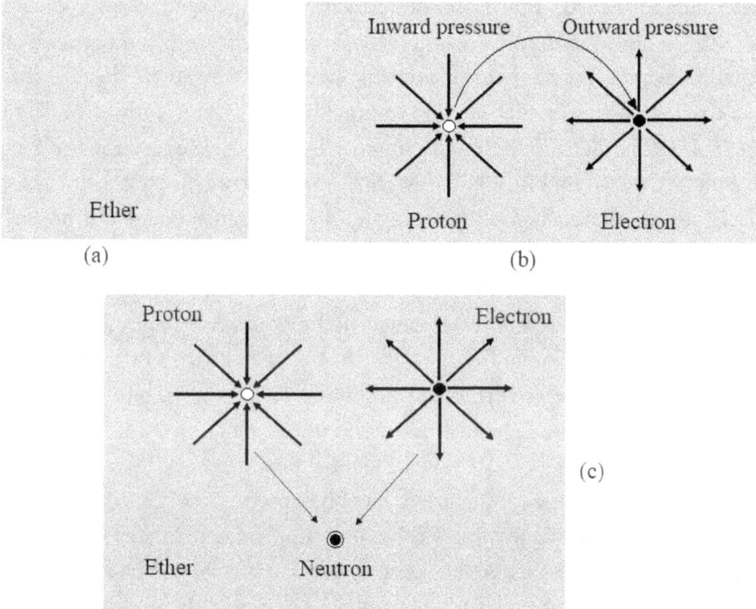

Figure 7.2.1. Production of electrons, protons and neutrons from ether.

If we want to think about this universe independently once again, we have to start from the twenty-first century. In fact, it would seem ridiculous in the current situation. But it is wrong to restrain oneself from expressing one's thoughts until someone else proves them wrong or until one convinces oneself that one's thoughts are wrong. At least one piece of evidence that is important from this point of view is that increasing the frequency of electromagnetic waves increases the magnetic force they apply, and this may change the way humans view the universe. First it is necessary to understand what fundamental particles are i.e., electrons, protons and neutrons. For that reason, to get involved in the history of their discovery and the ideas that scientists have about them at the moment is to move away from the ability of our intellect to imagine the universe from the experiences and observations that have come so far. Therefore, it will be right to give importance to what we feel without going into anyone's imagination. Fundamental particles are of two types namely electrically charged particles and neutral particles. We assume that an electron has a negative charge and creates a negative electric field around it. It is also assumed that a proton has a positive charge and creates a positive electric field around it. Both the above charges have the same effect but the fact is that we do not know what the charge is and cannot experience it. We don't know what's really going on with electrons and protons. Unlike charged particles i.e., electrons and protons attract and like charged particles i.e., electron-electron or proton-proton repel so we assume they have a charge because they must have something. We can experience attraction and repulsion between fundamental particles, but no one can say for sure that their charge is responsible for this. In fact, no one can say why there is attraction and repulsion between them. There must be some reason for this and it is not known and why scientists have not been able to find it or why they did not think it necessary to find it. It should have been revised only when attraction and repulsion were realized in elementary particles. But as a compromise they seem to have been released on the assumption that they should be charged, and on that basis the onward journey seems to have started. Mathematical equations can be written about the extent to which these particles attract and repel each other, but these equations cannot explain why this is happening. For this, one should be able to look at this incident from the right point of view and guess who exactly could be responsible for it. And one must also understand whether that reason is compatible with other happenings. Fundamental particles, ie electrons and protons, do not know whether they have a charge on them, but it is known that they have an electrical field around them. When a charged particle oscillates or vibrates, an electromagnetic wave is produced and propagates. It carries no charge with it as it travels, but when it interacts with another charged particle, it applies a force on it. This means that whatever is transmitted in the form of a wave is called a field. At present we understand that there are two types of fields in such a wave, namely the electric field and the magnetic field. This means that the existence of an electric field is a fact, not a fiction. Also, the same electric field is around the charged particle. That is, it is not known whether an electron or a proton has a charge on it, but there is a field around them, and that is the electric field. We have already discussed about magnetic field which can be an effect of asymmetric electric field. Now if it is a reality that the field exists, what is it? We just mentioned that the field propagates through the wave, which means that there must be a medium for the wave to form, which scientists called the ether. We will discuss the discovery of Ether in detail later but for the present situation we have to accept the existence of Ether because it makes no sense to say that a wave can be created without any medium. And if the wave is created without any media, then we will not be able to make any reasoning about this topic. We may record observations but not understand the true causes of their effects. Because there is a chain of thought and if it is broken, no conclusion can be reached. If we know that a wave exists, we know its frequency, wavelength, and speed, then the medium responsible for its production must exist, and if we want to find it, we must be able to predict its proper properties. If the assumptions about the property are wrong, it is difficult to find the thing. But one thing is true that the medium used to form the wave must be elastic to that field. Scientists have named this medium the luminiferous ether which should cover the entire universe. Now the question is how a spherically symmetric electric field can be formed in such a medium which is forming around the fundamental particles. Field means the pressure in that medium and spherically symmetric means the spherically symmetric pressure. Now the only thing we take for granted in this universe is ether which is perfectly elastic and pervades the entire universe. Now even if this assumption of ether solves all questions, why does this ether even exist? I don't think we will ever get the answer. So, what is the point of tracking this world if we are only going to be disappointed in the end. But man is a thinking animal, his intellect never lets him rest. Now the question before us is what can create spherically symmetric pressure? Because in the beginning there was ether which is perfectly elastic and nothing else in this universe.

Suppose a portion of ether is removed and placed elsewhere, either it must mix with ether and unite to form uniform ether like the first, or it must remain in the same place without mixing or uniting. If left in place, outward symmetric pressure may build up on its side. Also, there should be a cavity at the place from where that portion was removed and that cavity should also remain because the portion of ether removed from there which is placed elsewhere is the same, it has not been mixed. Now inward pressure should build up around this cavity. That means two points can be created in the ether, around one there is inward spherical pressure called positive spherical electric field and on the other there is outward pressure called negative spherical electric field. Now these two points are two different kinds of fundamental particles, aren't they? First proton and second electron. Now the effect of the field generated around both the points is going to be the same. What will they do now? Will they stay in place? Or will they try to get closer to each other? Ether is a perfect elastic medium and it is natural that it will always try to reduce the pressure created in it. There is no need to make a separate effort for that. There is no reason to have feelings. If the ether tries to remain neutral, these two points will try to move closer together. This process will also be natural. Now the question is, will both the points approach each other with equal speed? Or will they not have the same speed? Of course, the place where the part of ether is placed can move faster than the cavity of ether and from this it can be determined who is light and who is heavy. Now the point on which the ether part is placed can be called an electron and the point on which there is a cavity can be called a proton. Perhaps this is why we have come to understand that the masses of electrons and protons are different. This means there is no mass and no charge. When these two points i.e., electron and proton come close to each other, the pressure becomes neutral i.e., the electric field becomes zero. Now the question is, will it form a uniform ether after mixing them back together? Won't those two points merge into each other and stay the same? If those two points, i.e., electron and proton, do not mix with each other and continue to exist, then the pressure of both of them in the ether, i.e., the electric field, will be zero but they will continue to exist as one point or one particle. Maybe it's a neutron produced. If so, its mass must be equal to the sum of the masses of the electron and the proton. But perhaps because they overlap each other, their total mass may fall short of that sum. Now it will be difficult to move this neutral particle through the ether. Also, when two electrons are brought closer to each other the pressure in the ether will increase, so naturally there must be a repulsion between them to reduce the pressure. Similarly, proton-proton repulsion will be formed. Now all the electrons found in the universe have the same field strength and also all the protons have the same field strength. It can also mean that if an electron or negatively charged particle is produced that is less or greater than the strength of the electric field of the detected electron, a proton or positively charged particle of the same strength is produced and such particles mix back into the ether and disappear. But why protons and electrons with a certain field strength do not merge back into the ether and disappear after being created, cannot be answered yet. It is currently impossible to reason about how such a large number of electrons and protons could have been created. Also, one should be able to reason about how many types of fields can exist in the ether. Of course, only one type of pressure means only one type of field can exist and that is the electric field and the force produced by it is the electric force. Now the magnetic field is an effect of the asymmetric electric field so there is no reason for the magnetic field to exist independently. Then we see the existence of another field at large scale which is the gravitational field which seems to exist through the gravitational force. In fact, if the field existing in the ether is assumed to be electric, then no other field can exist in the ether. This means that magnetic field or gravitational field cannot exist in the either. The puzzle of what gravitational force is or what its true nature is still not solved.

No one seems to have reasoned about exactly how many types of fields can exist in the ether or in the universe itself. It is miraculous that no one seems to object to the fact that any number of fields can exist in the universe. In general, we have come to accept the existence of three types of fields in the universe, electric, magnetic and gravitational. If the magnetic field is an effect of an asymmetric electric field, it is not yet understood what the effect of the gravitational field is. Only the existence of an electric field holds true. But no matter what anyone claims, gravitational waves do not appear to exist, so it is true that the ether does not support the gravitational field. Gravitational force may be an attempt to bring together disturbances in the ether. So, this force will be an attempt to bring together all the disturbances in the ether whether it is inward pressure or outward pressure or some other kind. Electric field waves are also disturbances in the ether, so they must also exhibit a gravitational force, and that is the deflection of light rays in a gravitational field, the effect that catapulted Einstein into the limelight. Of course, Einstein's predictions about this are based on the General Theory of Relativity. If the gravitational force is an attempt to bring together the disturbances in the ether, two light rays traveling in close proximity should experience a gravitational force between them. Of course, since the speed of light is very high, it will be very difficult to find this gravitational force, but it is sure to be a fact. It will be interesting to see what the General Theory of Relativity will explain about this if one proves its existence experimentally. Although it is still not clear what exactly is gravitational force, it is necessary to think about it in line with the above. Another important thing about the ether is that if an object is stationary or in uniform motion in the ether, it stays that way unless an external force is applied to the object. This mystery is probably due to the fact that ether is a perfect elastic medium. When an object is traveling with uniform speed in the ether, at that time, if the resistance created by the ether in front of the object is equal to the support created to the elimination of the disturbances in the ether behind the object, then the object will continue to move forward through the ether with the same speed. This is called Newton's first law. This can only happen because ether is a perfect elastic medium. If we consider the spin motion of fundamental particles, the electron can enter easily into spin motion because at its point there is an additional part of ether but at the place of proton there is cavity of ether so that cavity may not easily hold. Similarly, may happen with neutron. Of course, these are only guesses, and not much can be predicted until we fully understand what the fundamental particles are from the above perspectives.

7.3 Emission and absorption of light rays

The visible light spectrum is the section of the electromagnetic spectrum that can be seen by the human eye. More simply, this range of wavelengths is called visible light.

As shown in figure 7.3.1, the electromagnetic waves falling in a certain band can be detected by our eyes, so that band is called the visible band of the spectrum, and the electromagnetic waves falling in this band are called light waves.

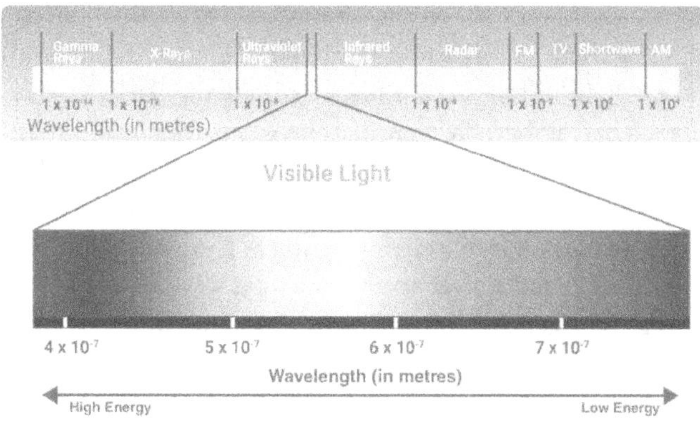

Figure 7.3.1. Visible band of spectrum.

Light is of great importance in human life as well as in the life of other animals. Because it is because of this light that we can see and feel everything around us. Light rays are generally believed to be produced by the transition of electrons in atoms. When an electron in an atom jumps from a higher energy state to a lower energy state or from an outer orbit to an inner orbit, the difference in energy is converted into electromagnetic waves. Suppose the energy of the electron in the upper energy state is E1 and that in the lower state is E2 and the difference between them is E1-E2 = hf then a wave of frequency f is generated, where h is Planck's constant. The funny thing here is that no one knows and no one talks about how that energy difference is converted into an electromagnetic wave. So, it seems that the research is going on according to our convenience. We are using electromagnetic waves in the frequency range of 30 KHz and 300 GHz for communication. To produce them, the length of the antenna is usually proportional to the wavelength, and an alternating current of the frequency of the wave to be produced is produced in the antenna. Therefore, a wave of that frequency is generated from that antenna and propagated in space at the speed of light. The message we want to send is encoded on that wave and the message is decoded by receiving that wave again. This means that we know exactly how electromagnetic waves are created. Light is also an electromagnetic wave, so, it does not explain why the same thing does not apply to the creation of light. If light of frequency f is emitted from an atom, it means that the corresponding electron is expected to vibrate in the same place at frequency f after jumping into that atom. We have seen that a wave of light emitted from a sodium atom has 650 crests or troughs. Although 650 is an approximate number, at least around that number the electron must have vibrated in one place due to the jump of the electron that created this wave, but no one seems to think about this. The reason for that is the wrong concept that either electrons move in orbits in atoms or express themselves as waves. It started at the beginning of the 20th century when Rutherford demonstrated that the atom has a heavy positively charged center and lighter negatively charged particles, called electrons, are distributed around the center. This raises the question why electrons do not fall into the nucleus. Perhaps they revolve around the nucleus in which the centripetal force and centrifugal force must balance, and this concept was derived from planetary motion. Then the problem arises that if the electrons continue to orbit, their energy should decrease and eventually fall into the nucleus, but this cannot be observed because the atoms are stable. On this, Bohr, while presenting his theory of the hydrogen atom, proposed that the atom must have some stationary orbits in which the electrons do not emit their energy as they move. Only when it jumps from one stationary orbit to another, its energy difference is released in the form of electromagnetic wave. Is it on the minds of electrons not to radiate energy while moving in stationary orbits? This is a pure compromise. But this arrangement made it possible to analyze the spectra of hydrogen by coincidence. But Bohr's theory is not applicable for large atoms. Further, the contributions of Planck and Einstein led to the emergence of quantum mechanics in the early 20th century. Einstein, in explaining the photoelectric effect, proposed that light can behave as a wave despite being a particle in which the light's energy converges to a point called 'light quanta', now called photons, which have energy hf, where h is Planck's constant and f is the frequency of that light. Light is a wave. But a question arises that if a light wave emitted from a sodium atom has 650 wavelengths, then the energy of that complete wave must be equal to one photon which is a point. Then where is the location of this photon in a wave of such length? Many scientists also believe that when a light wave is emitted from an atom, trillions of photons are emitted, and when one of these photons is detected by a detector, all the photons combine to show the effect of one photon because one wave is equal to one photon. That means any compromise is being made to support the photon, knowingly and unknowingly. Such a situation has arisen that everyone has lost their sanity. I don't know how everyone came under the influence of Einstein. This is just the beginning and the havoc is yet to come. Einstein explained the photoelectric effect by assuming that light is a wave and can also be expressed as a particle. Naturally, other implications began to be sought, and as a result, de Broglie proposed that a particle in motion could be expressed as a wave, which was the reverse of Einstein's proposal. That is, if a wave can be expressed as a particle, then a particle in motion can also be expressed as a wave. De Broglie also stated its wavelength. This led people to think that if a burning incense is rotated in a circle, the circle that appears is the de Broglie wave. This means that man has always expected miracles. In fact, that circle is our illusion and we watch TV or mobile every day on that illusion. That circle has nothing to do with the de Broglie wave. One such question is being discussed that what will be De Broglie wave? Because light wave is electromagnetic i.e., vector wave and De Broglie did not comment anything about this. Many say that wave is a scalar wave but no one answers why. In fact, learned people keep silent in times of trouble. After this, the next journey started with the assumption that if a particle is in motion at the microscopic level, it manifests itself as a wave, and this led to the development of 'Quantum mechanics.' Therefore, rather than whether electrons move in orbits or not, they react as waves in atoms was assumed. An attempt was made to discover the properties of atoms from this concept. This made understanding the atom more complicated. The fact is that we still don't know the structure of the atom, but we are researching black holes on a large scale and subparticles on a small scale. We have failed to understand the structure of the atoms we are made of. No one feels anything strange about it. That's why we think our research is convenient.

In 1925, Uhlenbeck and Goudschmidt proposed that the electron has a spin motion that gives it intrinsic magnetic momentum. But there is no discussion about whether spin can contribute to things other than its magnetic properties. It is mentioned that it is an intrinsic property of electrons and from that point of view, the distribution of paired and unpaired electrons in atoms and the creation of magnetic properties have been researched. But spin is not an intrinsic property of electrons, it can be proved from the experiment discussed earlier and also due to the spin motion, the electron does not need orbital motion to stabilize in the atom, but the reality is coming forward, so that the substance can get a solid shape. And so, if an electromagnetic wave with 650 crests or troughs emanates from an atom at a time, the corresponding electron must vibrate 650 times in one position. This is possible only if the electrons are not moving in orbit or it is not expressed as a wave. If we think in this direction, we will understand that when the electron jumps from a high energy state to a low energy state, how the energy difference is converted into an electromagnetic wave.

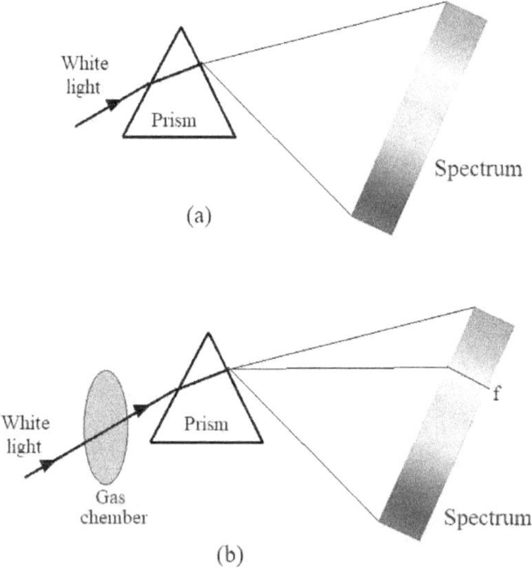

Figure 7.3.2. (a) Spectrum after white light passes through a prism, (b) Spectrum after absorption of a light wave of frequency f by atoms in a gas.

Another property of atoms is absorption and emission of spectra. Figure 7 .3.2. (a) shown is the spectrum seen after passing white light through a prism that includes all the colors. If an atom emits a light wave of frequency f, then if light waves of all frequencies pass through a gas of such atoms, the light of frequency f is absorbed and the rest of the light waves come straight out of the gas and a spectrum will appear as shown in Figure 7.3.2 (b). It is understood that when the atoms in a gas are in their ground state, when light waves of all frequencies pass through them, the electrons, emitting the light of frequency f when jumping from a higher energy state to a lower energy state, absorb the photons from the light waves of frequency f and goes into excited state. Therefore, those photons in the spectrum of light coming out of the gas will appear absent. Therefore, the line of that frequency will appear black in the spectrum of the light coming out. From this, a question arises that after the electron absorbs a photon of this frequency f and goes into an excited state, that atom will not be able to absorb another photon of the same frequency until it jumps to the ground state. And once it jumps, the same photon will be released again. The number of gas atoms in that gas chamber is limited and the light falling on that gas is continuous. So once all the atoms in the gas absorb photons of frequency f from that light, they will go into excited state and remain in the same state. Therefore, the photons of the frequency f that come later in that light will pass straight through the gas and no black line will be seen in the spectrum of the light that comes out after that. If the excited atoms return to the ground state immediately, they will re-emit the photons. Therefore, no black line will be seen in the spectrum of the light that comes out. Suppose the direction of the photons re-emitted by the absorbed atoms is random, but should be found. But there is no such discussion. This means that we don't know exactly what happens when an atom absorbs an electromagnetic wave. If there is a phase difference of 180 degrees between the absorbed wave and the corresponding electron that will vibrate and the wave created due to its vibration, then it is necessary to think from the point of view that maybe no wave will come out of that atom.

7.4 Summery

1. A medium is required for field and wave to exist, which has been named ether and attempts have been made to find it but could nit be successful. This does not mean that ether does not exist. If ignorance fails to properly assume the properties of a thing, it is difficult to find such a thing.

2. A field is the pressure in its associated medium. A spherically symmetric electric field is a spherically symmetric pressure which can be of two types, inward and outward. The points around which such pressures exist may be referred to as fundamental particles. The point around which there is inward pressure is the proton and the point around which there is outward pressure is the electron. If a small portion of that medium is removed from one place and placed at another, pressure will build up around both points. An inward pressure will build up around the place from where the particle was removed, which can be called a proton, and an outward pressure will build up around the place where the part was deposited, called an electron. So, it cannot be said that there is any such thing as charge.

3. The universal medium which we call ether will always try to remain neutral and it will create electric field waves and it will cause to create attraction between electron-proton and repulsion between electron-electron and proton-proton. A neutron can form as a particle if the electron and proton recombine and their combined existence exists.

4. The fields of electrons and protons i.e., the pressure around them has a certain value. Why this is so remains to be elucidated.

5. A single field can exist in the ether. This means only one kind of field wave will exist and if it is called electric field wave then magnetic field wave and gravitational field wave will not exist.